Neural Stimulation

Volume I

Editors

Joel B. Myklebust, Ph.D.
Associate Professor
Department of Neurosurgery
Medical College of Wisconsin
Milwaukee, Wisconsin

Joseph F. Cusick, M.D.
Professor
Department of Neurosurgery
Medical College of Wisconsin
Milwaukee, Wisconsin

Anthony Sances, Jr., Ph.D.
Professor and Chairman
Biomedical Engineering Department
Medical College of Wisconsin
Milwaukee, Wisconsin

Sanford J. Larson, M.D., Ph.D.
Professor and Chairman
Department of Neurosurgery
Medical College of Wisconsin
Milwaukee, Wisconsin

CRC Press, Inc.
Boca Raton, Florida

Library of Congress Cataloging in Publication Data
Main entry under title:

Neural stimulation.

Bibliography: p.
Includes index.
1. Neural stimulation — Therapeutic use.
2. Transcutaneous electrical nerve stimulation.
I. Myklebust, Joel B. II. title.
RC350.N48N48 1985 616.8′0645 84-23256
ISBN 0-8493-5253-3 (v.1)
ISBN 0-8493-5254-1 (v.2)

Direct all inquiries to CRC Press, Inc., 2000 Corporate Blvd., N.W., Boca Raton, Florida, 33431.

International Standard Book Number 0-8493-5253-3 (Vol. 1)
International Standard Book Number 0-8493-5254-1 (Vol. 2)

Library of Congress Card Number 84-23256
Printed in the United States

PREFACE

Neural stimulation encompasses a large and diverse body of experimental and clinical investigation. It is the goal of this text to coalesce the current concepts and clinical implementations of electrical stimulation of the nervous system. The first six chapters deal with theoretical and laboratory studies while the latter chapters discuss particular applications of neural stimulation.

Theoretical considerations regarding the distribution of electric fields in biologic media are dealt with by Plonsey. Jarzembski discusses the theoretical foundation for the measurement of current distribution secondary to electrical stimulation. Swiontek and Sances present experimentally derived current density plots and indicate the necessary adaptations and practical considerations for making these measurements. The use of the evoked potential as a means of investigating the mechanisms of neural stimulation is discussed by Myklebust and Cusick. Zuperku illustrates the laboratory use of electrical stimulation for the study of the central nervous system.

Perhaps the most common application for neural stimulation is the amelioration of pain. This stimulation may be applied at peripheral, spinal or central levels. Spinal stimulation for pain relief is reviewed by Maiman, Larson, and Sances, while Abram discusses peripheral stimulation (TNK), and Groth presents the findings with deep brain stimulation. Movement disorders have similarly been treated with stimulation at various levels of the central nervous system. Spinal stimulation for spasticity is extensively reveiwed by Sherwood. Gottlieb and Myklebust present the findings regarding the effect of cerebellar stimulation upon spinal reflexes. Harris discusses methods of assessing in spasticity with particular attention to changes with cerebellar stimulation.

It is well known that electrical stimulation of brain structures can modify hehavior. One of the pioneers in this area and his colleagues present a relatively new aspect of stimulation for behavioral disorders (Heath, Walker, Dempesy, and Franklin). Cranial electrotherapy stimulation is discussed by Smith.

One of the most exciting areas of investigation is in peripheral neuromuscular stimulation. The restoration of function to paralyzed limbs has become a realizable goal. McNeal and Bowman review the state of this important work. If any application of electrical stimulation can be termed clinically established, it is the use of electric fields for bone growth and healing. The current methods in this area are presented by Esterhai.

These volumes provide a significant overview of the current state of neural stimulation which we hope will be of considerable value to interested researchers and clinicians.

Joel B. Myklebust
Joseph F. Cusick
Anthony Sances, Jr.
Sanford J. Larson

THE EDITORS

Joel B. Myklebust, Ph.D. is a biomedical engineer in the Neuroscience Laboratory of the Veterans Administration Medical Center at Wood, Wisconsin. He received his B.S. degree from the University of Iowa in 1971 and his M.S. degree from the University of Rochester in 1972, both in electrical engineering. He received his Ph.D. degree in Biomedical Engineering from Marquette University of Milwaukee, Wisconsin, in June 1981. In July, 1981, he was appointed Assistant Clinical Professor of Biomedical Engineering in the Department of Neurosurgery at the Medical College of Wisconsin as well as Adjunct Assistant Professor of Biomedical Engineering at Marquette University, Milwaukee. In 1984, he was appointed Associate Professor of Biomedical Engineering in the Department of Neurosurgery at the Medical College of Wisconsin. His research interests include the therapeutic applications of electrical current to the nervous system and the recording of neuroelectric potentials for diagnostic purposes, and biomechanics.

Joseph F. Cusick, M.D., was born on May 10, 1939, in Binghamton, N.Y. He received the B.S. degree from Fordham University, Bronx, N.Y., in 1961 and the M.D. degree from Georgetown University School of Medicine, Washington, D.C., in 1965. He completed his residency in neurosurgery at the University of Illinois, Chicago, Ill., in 1972. He is presently Professor of Neurosurgery at the Medical College of Wisconsin, Milwaukee, Wisconsin, and Staff Neurosurgeon at the Veterans Administration Medical Center, Wood, Wisconsin. He is a member of the American Association of Neurological Surgeons, American College of Surgeons, Congress of Neurological Surgeons, International Society of Pituitary Surgeons, Research Society of Neurological Surgeons, and Cervical Spine Research Society. He received certification from the American Board of Neurological Surgeons in April, 1975.

Anthony Sances, Jr., Ph.D., has been Professor and Chairman of Biomedical Engineering at Marquette University and the Medical College of Wisconsin, Milwaukee, Wisconsin, since 1964. He is also Professor of Electrical Engineering and Professor in the Department of Neurosurgery. For the last 20 years he has been involved in the investigation of the effects of electrical currents upon animals and humans, and reconstruction of electrical injuries. Dr. Sances is also conducting studies to evaluate the biomechanical mechanisms of head and spinal cord injury. He has published 4 books and more than 200 scientific articles. He serves as a consultant to the National Institutes of Health, the National Science Foundation, and various industrial organizations. He is a registered professional engineer, a certified clinical engineer, and belongs to 24 national or international medical and engineering societies.

Sanford J. Larson, M.D., Ph.D., was born in Chicago, Illinois, on April 9, 1929. He received the B.A. degree from Wheaton College, Wheaton, Illinois, in 1950 and both the M.D. and Ph.D. degrees from Northwestern University, Evanston, Illinois, in 1954 and 1962, respectively.

After serving as an intern at Passavant Memorial Hospital, Chicago, Illinois, from 1954 to 1955, he took his residency in the Northwestern University Neurosurgical Residency Program from 1955 to 1957, and from 1959 to 1961, a period interrupted by 2 years of service in the Air Force. In 1961, he was awarded a Postdoctoral Research Fellowship and also served as Clinical Assistant in Surgery at the Northwestern University School of Medicine. In 1962, he was appointed Director of Neurosurgical Education at Cook County Hospital, Chicago, Illinois. In 1963, he was appointed Associate Professor and Chairman of the Department of Neurosurgery, Marquette School of Medicine (now Medical College of Wisconsin), Milwaukee, and is currently Professor and Chairman of that Department.

CONTRIBUTORS

Stephen E. Abram, M.D.
Professor of Anesthesiology
Medical College of Wisconsin
Milwaukee, Wisconsin

Bruce R. Bowman, Ph.D.
Vice President
Med Tel, Inc.
Edina, Minnesota

Colby W. Dempesy, Ph.D.
Research Professor in Neurosurgery
Tulane University School of Medicine
New Orleans, Louisiana

John L. Esterhai, Jr., M.D.
Assistant Professor
Department of Orthopaedic Surgery
University of Pennsylvania School of Medicine
Philadelphia, Pennsylvania

Dennis E. Franklin, M.D.
Assistant Professor of Psychiatry and Neurology
Tulane University School of Medicine
New Orleans, Louisiana

Gerald L. Gottlieb, Ph.D.
Professor of Physiology
Rush Medical College
Chicago, Illinois

Karl Groth, Ph.D.
Product Planning Manager
Metronics, Inc.
Minneapolis, Minnesota

Gerald F. Harris, Ph.D.
Director, Biomedical Engineering
Shriners Hospital
Chicago, Illinois

Robert G. Heath, M.D.
Professor of Psychiatry and Neurology
Tulane University School of Medicine
New Orleans, Louisiana

Francis A. Hopp, Ph.D., M.S.
Department of Anesthesiology
Medical College of Wisconsin
Biomedical Engineer
V.A. Medical Center
Milwaukee, Wisconsin

W. B. Jarzembski, Ph.D., P.E.
Professor of Biomedical Engineering
Texas Tech University School of Medicine
Lubbock, Texas

Dennis J. Maiman, M.D.
Assistant Professor of Neurosurgery
Medical College of Wisconsin
Milwaukee, Wisconsin

Donald R. McNeal, Ph.D.
Co-Director
Rancho Los Amigos Rehabilitation Engineering Center
Downey, California

Barbara M. Myklebust, M.S.
Department of Physiology
Rush Medical Center
Chicago, Illinois

Robert Plonsey, Ph.D.
Professor of Biomedical Engineering
Duke University
Durham, North Carolina

Arthur Sherwood, Ph.D.
Department of Clinical Neurophysiology
Institute of Rehabilitation and Research
Houston, Texas

Ray B. Smith, Ph.D.
Vice President for Science Affairs
Neuro Systems, Inc.
Garland, Texas

Thomas J. Swiontek, Ph.D.
Assistant Professor of Electrical Engineering
Milwaukee School of Engineering
Milwaukee, Wisconsin

Cedric F. Walker, Ph.D.
Associate Professor of Biomedical Engineering
Tulane University School of Medicine
New Orleans, Louisiana

Edward J. Zuperku, Ph.D.
Associate Professor of Biomedical Engineering
Department of Anesthesiology
Medical College of Wisconsin
Biomedical Engineer/Research Service
V.A. Medical Center
Milwaukee, Wisconsin

TABLE OF CONTENTS

Volume I

Volume II

Chapter 1

INTRODUCTION

J. Cusick and J. Myklebust

Electrical stimulation for manipulation of the nervous system has frequently been greeted with enthusiastic and sometimes uncritical acceptance. The resulting high expectations coupled with the relative technical ease of implementations have led to widespread use of neural stimulation without a complete definition of the mechanisms of action. The marginal functional gains obtained by many patients and the inevitable complications associated with innovative procedures have produced skepticism and conflict. This sequence of events in which an aggressive clinical application of certain stimulation methods has been subsequently temperated and clarified by follow-up laboratory studies has occasionally fostered a dichotomy of opinions between laboratory and clinical investigators. Such differences in thought are not usually the result of a clear conflict of experiences, but rather the result of poor communication between laboratory and clinical researchers. Later analysis, in these situations, frequently shows that the two approaches complement each other and clarify the value of various stimulation modalities.

An effort has been made in this text to correlate the dialogue between the clinical and laboratory experiences. The initial chapters describe laboratory studies of electrical stimulation with special emphasis on electrical field modeling and current density measurements. These studies address the physiologic and biophysical basis for the neurologic changes encountered with electrical stimulation of the nervous system. Later chapters present the clinical experience with neural stimulation and suggest areas for further studies, both in the clinic and in the laboratory.

The modern history of electrophysiology is characterized by the shift in emphasis from macroscopic to microscopic methods. Fritsch and Hitzig pioneered the use of electrical stimulation for the functional mapping of the cerebral cortex.[43] Currently prevalent intracellular methods have produced information on the characteristics of cell membranes and have resulted in improved understanding of the neuron and synapse as integral parts of the complex processes of perception and behavior.

Clinical stimulation of the nervous system is nearly always macroscopic and extracellular, and a complete definition of the mechanisms of action requires knowledge of the distribution of currents and the conditions by which the currents cause activation of cells or fibers. Although a comprehensive review of neurophysiology is beyond the scope of this chapter,[91,104] electrical stimulation may produce a wide variety of effects including general activation or inactivation of neural tissues, activation of excitatory and inhibitory systems, and alterations in biochemical concentrations.

This diversity of effect is reflected in the range of clinical applications of neural stimulation. Stimulation is used in an attempt to replace functions lost by injury or disease, as in sensorineural prostheses, in efforts to prevent the transmission of unwanted information, as in stimulation for pain, or to promote healing, as in stimulation for bone growth.

The most ambitious and complex of these applications is the effort to use electrical stimulation for direct functional activation of the nervous system. Progress in this area is dependent upon the development of understanding of both the complex integration of the nervous system and the means by which electrical stimulation can mimic this integration. One of the most important side effects of the clinical application of functional neural stimulation is the information obtained concerning human neurophysiology, for the majority of present basic neurophysiologic knowledge has been obtained from human pathology and animal studies. Therefore, neural stimulation offers unique opportunities to learn about the normal function of the human nervous system. This situation again emphasizes the symbiotic relationship between laboratory study and clinical application which is essential to progress in this field.

NEURAL DAMAGE

In determining the legitimate clinical application for an implantable electrode system, definition of the complications or risks associated with the various devices is as important as the careful analysis of the beneficial effects. It is very reasonable to agree with the assumption that imposition of a foreign body (implanted electrode system) will result in tissue alterations both from the trauma of the surgical procedure and changes induced by reactions to the electrode material. In conjunction with these mechanical changes, these same tissues will also be exposed to localized passage of unnatural electrical currents, and therefore, the questions of irreversible neurologic damage affecting both the functional capabilities of the patient, the effects of these tissue changes at the electrode-tissue interface, or the anticipated character of the electrical stimulation become important considerations.

The effects from surgical implantation will vary relative to the region involved, the character of device, the recipient's medical condition, and the technique of implantation. It is generally understood that any surgical intervention will alter normal anatomic spaces and local tissue character and that these alterations can frequently be minimized by careful attention to surgical techniques. As important as surgical technique is the need for the device to be compatible with its new environment with special consideration to restricting the size to prevent tissue compression or distortion and to use materials of low tissue reactivity. In this latter regard, it is generally accepted that platinum or platinum alloys are the preferred electrode material because of their apparently innocuous local electrochemical reactions and resistance to corrosion.[20-22] This conclusion is not meant to convey that the possibility of adverse electrochemical reactions at the electrode-tissue interface could not occur, but that such occasions appear to be of limited scope. It has been suggested that capacitive electrodes should be used to eliminate electrolytic reaction at the electrode.[3,53] However, platinum electrodes produce minimal reactions and the added expense may not be justified. Recent animal studies of the matrix for surface electrode arrays compared silastic and Dacron® mesh and noted that the Dacron® mesh developed less current resistance and less local tissue changes. The contribution of electrode size appears to be a selfevident situation in regards to local tissue changes with the character of the electrode designed to cause little compromise of the anatomic space to be implanted. In considering this aspect relative to depth electrode insertion, one can always expect a certain element of tissue damage resulting from the elctrode's passage through the tissue, and in the CNS, these invasions will elicit repairative changes with glial reactions and cavitation. Such histologic alterations are directly related to electrode size, with electrodes measuring 0.5 mm or less minimizing such tissue changes to the degree that little secondary effects on electrode function would be anticipated.

A natural concern in the use of neural stimulation is the possibility of tissue damage resulting from the electric currents. Studies have been directed toward the development of noninjurious electrodes, waveform, and parameters as well as the evaluation of systems in clinical use. Since the observation of electrolytic injury with unidirectional currents by Horsley and Clark,[58] it has been recognized that bidirectional current flow should be utilized in neural stimulation. This can be achieved with balanced biphasic waveforms,[78] capacitively coupled pulses, or with sinusoidal alternating currents.[88] The electrolytic changes, however, produced at the electrode-tissue interface are not symmetric even with balanced biphasic waveforms.[119] Consequently, the parameters of biphasic stimulation required to produce damage have been extensively investigated using various criteria including alteration of the threshold for neural activation, blood-brain barrier changes, pH changes, and light and electron microscopic histologic changes. Various measured electrical parameters have been suggested as most important in accounting for the observed changes.

Current, voltage, and resistance are directly related by Ohm's law.*

$$\text{Current} = \frac{\text{voltage}}{\text{resistance}}$$

Although the most important parameter for physiologic stimulation is usually the magnitude of the current, the other critical parameters for electrical damage to tissue have been suggested to be power,[90] energy,[105] and charge transfer.[99,123] The pertinent definition of these terms is that power is the product of current and voltage and represents the instantaneous energy applied, energy is the integral of power over time, and charge transfer is the integral of current over time indicating the magnitude of ionic flow induced. These parameters may be expressed in absolute terms or may be adjusted for the particular electrode parameters or electrode size in use. For example, 1 pulse per second might be expected to be less damaging than 100 pulses per second since the current is less sustained, and 1 pulse per second with a duration of 0.1 sec might be expected to be less injurious than 1 pulse per second with a duration of 0.5 sec. Consequently, power is usually expressed as an average for the time the current is actually applied, for example, if 100 pulses per second with each pulse lasting 0.25 msec is used, the current is actually flowing

$$100\ \frac{\text{pulses}}{\text{sec}} \times \frac{0.00025\ \text{sec}}{\text{pulse}} = 0.025$$

of the time. The power measured for a single pulse is multiplied by this factor (the duty cycle) to obtain the average power.

The charge transferred with each pulse is the area of the current waveform as a function of time. This parameter may also be averaged by the duty cycle. Another often used indicator is the cumulative total charge transferred. For a charge per pulse of 1 μC (1 C is 1 A for 1 sec), with 100 pulses per second, applied for 1 min, the total charge is

$$\frac{1\ \mu\text{C}}{\text{pulse}} \times \frac{100\ \text{pulses}}{\text{sec}} \times 60\ \text{sec} = 6000\ \mu\text{C}$$

These parameters are also usually adjusted for the size of the electrode, and in this respect, smaller electrodes which deliver a more intense stimulus to a smaller region might be expected to have a lower threshold for damage. Although theoretical and experimental data suggest that the electrode circumference is the important geometrical property,[44,94,105] the electrical parameters (current, power, and charge) are usually divided by the area of the electrodes in use.

Using the criterion of alteration in the blood-brain barrier, Mortimer et al.[90] found that capacitively coupled or bidirectional pulses with average power of 0.05 W/in.^{2} (0.0078 W/cm^2) were safe. Monophasic waveforms with average power of 0.003 W/in.^{2} (0.00046 W/cm^2) were also judged to be safe, and focusing on the delivered charge, Pudenz et al.[99] reported that the blood-brain barrier was not altered at 0.45 μC per pulse, but was affected at 1.0 μC per pulse. In contrast, Rowland et al.[103] found that bidirectional pulses of 20 μC per pulse with cumulative charges of up to 20 C did not produce damage. Larson et al.[76] applied capacitively coupled pulses of up to 10 mA with charge per phase of 2.5 μC and average power up to 0.4 W/in.^{2} (0.062 W/cm^2) without blood-brain barrier changes. Swiontek

* Since impedance is a frequency-dependent quantity, this relationship is more properly expressed in terms of the Laplace or Fourier transforms. However, since most neural stimulation is conducted with a fixed waveform, this relationship is sufficient. For clarity, resistance is sometimes referred to as "access resistance", the apparent resistance for the particular waveform and electrode.

et al.[113] reported that capacitively coupled pulses of up to 10 mA (2.5 μC per pulse) did not change the pH of the tissue under the electrode, and monophasic pulses of 2 to 3 mA (0.5 to 0.6 μC per pulse) began to increase the pH at the negative electrode with a corresponding decrease at the positive pole.

Histological change following electrode implantation has been attributed to both electrolytic and mechanical causes. Dobelle[35] applied 5-mA, 0.5-msec-duration, 50-Hz balanced biphasic pulses to the cat cortex for 660 min through 1-mm-diameter platinum ball electrodes with minimal histological change. Gilman et al.[45,46] found histological damage in the monkey cerebellum including loss of Purkinje cells and gliosis with stimulation at 2.4 μC per pulse. Larson et al.[76] found similar changes in the monkey, but attributed them to the mechanical compression caused by the electrodes. Brown et al.[19] found no histologic damage due to stimulation at levels of 0.5 μC per phase (7.4 μC/cm^2 per phase), but did find changes at higher levels. The effects were more pronounced and changes were found at lower stimulus levels with electron microscopy.

In long-term experimental studies, Larson et al.[77] found no physiologic or histologic change due to stimulation in chimpanzees with 5.5 to 6.5 mA (1.4 to 1.6 μC per pulse) for up to 3 months. After intervals of 7 to 36 weeks, Dauth[31] noted reactive fibrosis in monkeys with more change at the stimulating electrodes than at the control sites. They saw localized cortical damage, even with the control electrodes, consisting of decreased numbers of Purkinje cells and degeneration of molecular and granular cell layers, presumably due to direct compression. In animals undergoing continuous stimulation, changes in number as well as morphology of Purkinje cells were seen, and the severity of the damage was inversely proportional to the distance from the electrodes. In animals stimulated intermittently, little injury was seen away from the immediate area of the electrodes, and even in animals with severe cerebellar damage, there were no signs of neurologic injuries or changes in evoked potentials. These findings have resulted in the hypothesis that the histologic changes are due to products of electrolysis and pH changes.

Histological evaluation of neural tissue from patients that had been stimulated noted the consistent thickening of the meninges with encapsulation of the electrodes[30,77,102] and mechanical deformation of the tissue with cell loss immediately under the electrodes was routine. These findings led to the conclusion that in the majority of cases the effects were the result of mechanical causes. Little evidence of neurologic sequelae has been published.

Implantation of electrodes and stimulation of the nervous system produces histologic changes in neural tissue. However, it must be concluded that at normal clinical levels, the induced effects are slight relative to the pathologies in which these methods are applied. With proper selection of electrodes and parameters of stimulation and careful implantation, the character of neural injury due to stimulation relative to the clinical situation is generally not sufficient to preclude clinical application.

VISUAL AND AUDITORY PROSTHESES

In view of the broad-based character of the subject matter to which this text is dedicated, it is difficult to offer an in-depth review of all modalities of neural stimulation, and, in this regard, omissions in the text such as auditory and visual prostheses, bladder stimulation, and phrenic nerve pacing will be briefly reviewed in this introduction.

Implantable visual or auditory prosthetic systems may be placed over the appropriate region of the cerebral cortex[16,35,36,38] or at more peripheral sites,[34,90,110] and because of the difficulties in exposing the auditory cortex, located deep within the Sylvian fissure, auditory stimulation has primarily been applied to the cochlea or auditory nerve. The visual cortex, however, is more accessible and the efforts for artificial vision have been concentrated on cortical stimulation.

There are obvious advantages to encapsulating and implanting the stimulation system and using inductive coupling to transmit the stimulus parameters and this approach has been applied to both auditory and visual systems.[16-18,39,65,121] The inaccessibility of the electrodes for determining the delivered stimulus and the inability to refine the stimulation system as new information becomes available led others to develop a percutaneous electrode connector so that an external stimulation system could be used to develop the stimuli applied to the neural tissue.[71,89] The prosthetic systems consist of a transducer (a camera for the visual system and a microphone for the auditory system), a processor which converts the transducer output into a code suitable for stimulating the tissue, a system for relaying the stimulus to the electrodes, and the electrodes on the neural tissue.

Penfield, relying upon early investigations,[40,72] produced visual sensations with electrical stimulation of the cortex during resection of lesions under local anesthesia.[95,96] Although a few researchers suggested visual cortex stimulation for visual prosthesis,[23,73-75, 109] it was not until 1968 that Brindley and Lewin conducted their pioneering experiments in which 81 subdural electrodes were implanted.[16] Their report confirmed that small reproducible phosphenes (punctuate sensations of light) could be elicited in totally blind subjects. Phosphene effects have been reported to include ''clouds'', ''pinwheels'', and occasional chromatic sensations, as well as simple punctuate ''stars in the sky''.[40,72,95,96]

The direction of visual prosthesis research has been to implant an array of electrodes on the visual cortex. By stimulation of selected electrodes, images can be produced as sets of phosphenes. Consequently, the first task following the implantation of electrodes is the production of a correspondence mapping of visual space to the electrode array. Subsequent alterations of the stimulus parameters will allow variation of the perceived brightness. As previously noted, the key question for visual prosthesis is whether useful images can be transmitted by stimulation of appropriately selected sets of electrodes, and at the present time the ''state of the art'' is that an electrode array encoded as a 2×3 Braille cell permits a trained volunteer to read Braille 5 times faster than he could by using his fingers.[37] Although the facilitation of Braille reading does not justify the widespread application of implanted electrodes, it is clear that useful, reproducible images can be transmitted. These findings suggest that improved resolution from larger electrode arrays, increased understanding of visual electrophysiology, and refined stimulation systems may make a functional visual prosthesis a reality.

Direct stimulation of the acoustic nerve was accomplished in 1957 by Djourno and Eyries.[34] An electrode was temporarily implanted on the VIII nerve of a bilaterally deaf patient. The sounds produced by stimulation resembled a spinning roulette wheel or chirping crickets and varied with the frequency of stimulation. The rhythm of the sounds reportedly aided the patient in lip reading. Simmons implanted six electrodes in the modiolar VIII nerve tissue of bilaterally deaf volunteers.[110,111] Variations in pitch were related to the electrode position and stimulus frequency, and the loudness depended on stimulus amplitude. About the same time, House and Urban[60,61] and Michelson et al.[85-87] utilized single-channel electrode systems for stimulation in the cochlea. Although these devices did not produce recognizable speech, it was reported that patients could detect a ringing doorbell or telephone and read lips more easily. These devices remained in place for a number of years without substantial changes in the perceived sounds.

White estimates that approximately 100 cochlear implants are currently performed each year.[122] It is likely that a total of more than 600 have been implanted.[26,122] The earliest clinical systems utilized a single monopolar[49] or bipolar electrode[79-81] and an inductively coupled stimulation system. These single-channel devices are able to utilize periodicity pitch (the ability of auditory neurons to respond to a range of frequencies), but not place pitch (the coding of frequency selectivity by position within the cochlea). Although early results were equivocal,[2,4-9,11-13,59] improved results were obtained when the stimulated electrode was selected from an array,[56] or when an array of electrodes was driven as a unit.[122] Single-channel prostheses have produced an improvement in lip reading ability, intelligibility of speech, and an ability to recognize some environmental sounds.[32,62-64]

The limited spectral band width makes it unlikely that unaided speech comprehension can be achieved with the single-channel systems. Present research efforts indicate that a successful auditory prosthesis which provides speech comprehension requires a multichannel stimulation system that takes advantage of the tonotopic arrangement of the cochlea, and most recent research has proceeded in this direction.[28,29,55,57,66,82-84,117,120] Although there are approximately 30,000 fibers in the auditory nerve, it is feasible that as few as 8 to 12 stimulation channels may produce intelligible speech,[122] and remarkable results have been reported by Pialoux et al.[24,25,97,98] with 8- and 12-channel systems. In these studies, sounds are passed through band pass filters, and the amplitude of a filter output is used to frequency modulate a train of constant amplitude pulses applied to an electrode. It is reported that the understanding of a few words is achieved within days of surgery. Within 1 month, 50% of conversation is understood and most patients ultimately achieve greater than 80% comprehension. In some patients, an improvement in speech is observed.[98]

Optimal stimulation of the auditory system requires information concerning the effects of stimulus parameters and auditory neurophysiology. This information, in conjunction with recent developments in electronic technology, indicates that a functional auditory prosthesis is a realizable goal.

PHRENIC NERVE STIMULATION

Stimulation of the phrenic nerve for pacing the diaphragm has been applied clinically since 1948.[47-52] Most studies have been performed with bipolar platinum electrodes embedded in a silastic cuff to encircle the nerve. The stimulus is usually applied transcutaneously using the radiofrequency transmission system employed in other neural stimulation methods, but totally implantable units are in development.[106] Direct transcutaneous stimulation of the phrenic nerve is often used to screen candidates for the procedure.

Phrenic nerve stimulation is usually applied to chronic ventilatory insufficiency of central origin with complete dependence upon mechanical ventilation. Although this method might be expected to produce improved mobility, a lack of confidence and acceptance by the patient is not unusual. Disease states in which phrenic nerve stimulation has been used include encephalitis, brain stem infarction, brain stem tumor, posterior fossa tumor, Pickwickian syndrome, postcervical cordotomy ventilatory insufficiency, bulbar poliomyelitis, and traumatic injury of the upper cervical spinal cord.[51,100,101,108,118] It has been reported to be of little value in respiratory insufficiency of peripheral origin.[51,101] Histological studies have shown little damage to the phrenic nerve with clinical stimulation methods.[70] No changes were reported in the diaphram at low stimulation frequencies, but pathological alterations were observed at high frequencies.[27]

STIMULATION FOR BLADDER CONTROL

Considerable effort has been directed to the development of methods to electrically control micturition in spinal cord-injured patients.[1,10,14,15,33,41,54,67-69,92,107,112,115,116] Although several methods have been employed with varying degrees of success, routine clinical application remains in the future. Control of bladder function has been attempted by stimulation of the spinal cord, the sacral roots, the detrusor nerve and muscle, and the pelvic nerve.

Although stimulation of the spinal cord is nonspecific and often leads to undesirable responses, some favorable clinical results have been reported.[42,93] Stimulation of the more peripheral sites, however, might be expected to be more successful since greater stimulation specificity can be achieved. Problems including electrode migration, undesirable stimulation of competing mechanisms, and pain have led some to conclude that stimulation of the sacral roots is the most promising approach. None of these approaches has proven to be fully effective, and the frustrations encountered by all of these efforts are mainly the result of the inability to reproduce the complex integrated stimulation system required for micturition, for such a situation requires balanced control of the multiple components of the micturition system, which is beyond the capacity of present neural stimulation systems.

ACKNOWLEDGMENT

This research was supported in part by the Veterans Administration Medical Center Research Grant #1655-03P.

REFERENCES

1. **Alexander, S. and Rowan, D.,** Experimental evacuation of the bladder by means of implanted electrodes, *Br. J. Surg.*, 52, 808, 1965.
2. **Ballantyne, J. C., Evans, E. F., and Morrison, A. W.,** Electrical auditory stimulation in the management of profound hearing loss. Report to the Department of Health and Social Security on visits in October 1977 to centers in the U.S.A. involved in cochlear implant prostheses, *J. Laryngol. Otol.* (Suppl.), 1, 1, 1978.
3. **Berstein, J. J., Hench, L. L., Johnson, P. F., Dawson, W. W., and Hunter, G.,** Electrical stimulation of the cortex with tantalum pentoxide capacitive electrodes, in *Functional Electrical Stimulation: Application in Neurol Prostheses,* Hambrecht, F. T. and Reswich, J. B., Eds., Marcel Dekker, New York, 1977.
4. **Bilger, R. C. and Hopkinson, N. T.,** Hearing performance with the auditory prosthesis, *Ann. Otol. Rhinol. Laryngol. (Suppl.),* 86(3, Part 3: Suppl. 38), 76, 1977.
5. **Bilger, R.C. and Black, F. O.,** Auditory prostheses in perspective, *Ann. Otol. Rhinol. Laryngol.(Suppl.),* 86(3, Part 2: Suppl. 38), 3, 1977.
6. **Bilger, R. C.,** Electrical stimulation of the auditory nerve and auditory prostheses: a review of the literature, *Ann. Otol. Rhinol. Laryngol. (Suppl.),* 86(3, Part 2: Suppl. 38), 11, 1977.
7. **Bilger, R. C., Black, F. O., and Hopkinson, N. T.,** Research plan for evaluating subjects presently fitted with implanted auditory prostheses, *Ann. Otol. Rhinol. Laryngol. (Suppl.),* 86(3, Part 2: Suppl. 38), 21, 1977.
8. **Bilger, R. C., Stenson, N. R., and Payne, J. L.,** Subject acceptance of implanted auditory prosthesis, *Ann. Otol. Rhinol. Laryngol. (Suppl.),* 86(3, Part 2: Suppl. 38), 165, 1977.
9. **Bilger, R. C., Black, F. O., Hopkinson, N. T., and Myers, E. N.,** Implanted auditory prosthesis: an evaluation of subjects presently fitted with cochlear implants, *Trans. Am. Acad. Ophthalmol. Otolaryngol.,* 84(4, Part 1), ORL-677, 1977.
10. **Boyce, W., Lathan, J., and Hunt, L.,** Research related to the development of an artificial electrical stimulator for the paralyzed human bladder, *J. Urol.,* 91, 41, 1964.
11. **Black, F. O.** Present vestibular status of subjects implanted with auditory prostheses, *Ann. Otol. Rhinol. Laryngol. (Suppl.),* 86(3, Part 3: Suppl. 38), 49, 1977.
12. **Black, F. O. and Myers, E. N.,** Present otologic status of subjects implanted with auditory prostheses, *Ann. Otol. Rhinol. Laryngol. (Suppl.),* 86(3, Part 2: Suppl. 38), 25, 1977.
13. **Black, F. O.,** Effects of the auditory prosthesis on postural stability, *Ann. Otol. Rhinol. Laryngol (Suppl.),* 86(3, Part 2: Suppl. 38), 141, 1977.
14. **Bradley, W. E., Wittmers, L., Chou, S., and French, L.,** Use of a radio transmitter-receiver unit for the treatment of neurogenic bladders, *J. Neurosurg.,* 19, 782, 1962.
15. **Bradley, W. E.,** Experience with electronic stimulation of the micturition reflex function, in *Functional Electrical Stimulation: Applications in Neural Prostheses,* Hambrecht, F. T. and Reswick, J. B., Eds., Marcel Dekker, New York, 1977, 119.
16. **Brindley, G. S. and Lewin, W. S.,** The sensations produced by electrical stimulation of the visual cortex, *J. Physiol. (London),* 196, 487, 1968.
17. **Brindley, G. S., Donaldson, P. K., and Falconer, P.,** The extent of the region of occipital cortex that when stimulated gives phosphenes fixed in the visual field, *J. Physiol. (London),* 225, 57, 1972.

18. **Brindley, G. S.,** Sensory effects of electrical stimulation of the visual and paravisual cortex in man, in *Handbook of Sensory Physiology*, Vol. 7/B, Jung, R., Ed., Sprnger-Verlag, Basel, 1973, 583.
19. **Brown, W. J., Babb, T. L., Soper, H. V., Lieb, J. P., Ottino, C. A., and Crandall, P. H.,** Tissue reactions to long-term electrical stimulation of the cerebellum in monkeys, *J. Neurosurg.*, 47, 366, 1977.
20. **Brummer, S. B. and Turner, M. J.,** Electrical stimulation with Pt electrodes. I. A method for determination of "Real" electrode areas, *IEEE Trans. Biomed. Eng.*, 24(5), 436, 1977.
21. **Brummer, S. B. and Turner, M. J.,** Electrical stimulation with Pt electrodes. II. Estimation of maximum surface redox (theoretical non-gassing) limits, *IEEE Trans. Biomed. Eng.*, 24(5), 440, 1977.
22. **Brummer, S. B. and Turner, M. J.,** Electrochemical considerations for safe electrical stimulation of the nervous system with platinum electrodes, *IEEE Trans. Biomed. Eng.*, 24(1), 59, 1977.
23. **Button, J. and Putman, T.,** Visual responses to cortical stimulation in the blind, *J. Iowa Med. Soc.*, 52, 17, 1962.
24. **Chouard, C. H. and MacLeod, P.,** Implantation of multiple intracochlear electrodes for rehabilitation of total deafness: preliminary report, *Laryngoscope*, 86, 1743, 1976.
25. **Chouard, C.,** Multiple intracochlear electrodes for rehabilitation in total deafness, *Otolaryngol. Clin. N. Am.*, 11(1), 217, 1978.
26. **Chouard, C. H., Fugain, C., Donnadieu, G., Josset, P., and Lengrand, D.,** Electrical stimulation of the middle ear in total deafness: results in 375 cases (author's transl.), *Ann. Oto Laryngol. Chir. Cervico Fac.*, 99(1-2), 15, 1982.
27. **Ciesielski, T. E., Fukuda, Y., Glenn, W. W. L., Gorfien, J., Jeffery, K., and Hogan, J. F.,** Response of the diaphragm muscle to electrical stimulation of the phrenic nerve: a histochemical and ultrastructural study, *J. Neurosurg.*, 58, 92, 1983.
28. **Clark, G. M., Black, R., Dewhurst, D. J., Forster I. C., Patrick, J. F., and Tong, Y. C.,** A multiple-electrode hearing prosthesis for cochlear implantation in deaf patients, *Med. Prog. Technol.*, 5(3), 127, 1977.
29. **Clark, G. M., Black, R., Forster, I. C., Patrick, J. F., and Tong, Y. C.,** Design criteria of a multiple-electrode cochlear implant hearing prosthesis, *J. Acoust. Soc. Am.*, 63(2), 631, 1978.
30. **Cooper, I. S., Riklan, M., Amin, I., et al.,** A long-term follow-up study of cerebellar stimulation for the control of epilepsy, in *Cerebellar Stimulation in Man*, Cooper, I. S., Ed., Raven Press, New York, 1978, 19.
31. **Dauth, G. W., Defendini, R., Gilman, S., Tennyson, V. M., and Kremzner, L.,** Long-term surface stimulation of the cerebellum in the monkey, *Surg. Neurol.*, 7, 377, 1977.
32. **Dent, L. J.,** Vowel discrimination with the single-electrode cochlear implant: a pilot study, *Ann. Otol. Rhinol. Laryngol.*, 91, 41, 1982.
33. **Diokno, A. C., Davis, R., and Lapides, J.,** The effect of pelvic nerve stimulation on detrusor contraction, *Invest. Urol.*, 11, 178, 1973.
34. **Djourno, A. and Eyries, C.,** Prosthese auditive par excitation electrique a distance du nerf sensorial a l'aide d'un bobinage indus a demeure, *Presse Med.*, 65, 1417, 1957.
35. **Dobelle, W. H., Mladejovksy, M. G., Stensaas, S. S., and Smith, J. B.,** A prosthesis for the deaf based on cortical stimulation, *Ann. Otol. Rhinol. Laryngol.*, 82, 445, 1973.
36. **Dobelle, W. H., Mladejovksy, M. G., and Girvin, J. P.,** Artificial vision for the blind: electrical stimulation of visual cortex offers hope for a functional prosthesis, *Science*, 183, 440, 1974.
37. **Dobelle, W. H., Mladejovsky, M. G., Evans, J. R., Roberts, T. S., and Girvin, J. P.,** "Braille" reading by a blind volunteer by visual cortex stimulation, *Nature (London)*, 259, 111, 1976.
38. **Dobelle, W. H., Quest, D. O., Antunes, J. L., Roberts, T. S., and Girvin, J. P.,** Artificial vision for the blind by electrical stimulation of the visual cortex, *Neurosurgery*, 5(4), 521, 1979.
39. **Donaldson, P. E. K.,** Experimental visual prosthesis, *Proc. Inst. Elec. Eng. (London)*, 120, 281, 1973.
40. **Foerster, O.,** Beitrage zur Pathophysiologie der Sehbahn und der Sepsphare, *J. Physiol. Neurol. (Leipzig)*, 39, 463, 1929.
41. **Friedman, H., Nashold, B. S., and Senechal, P.,** Spinal cord stimulation and bladder function in normal and paraplegic animals, *J. Neurosurg.*, 36, 430, 1972.
42. **Friedman, H., Nashold, B. S., and Grimes, J.,** Electrical stimulation of the conus medullaris in the paraplegic, a five year review, in *Functional Electrical Stimulation: Applications in Neural Prostheses*, Hambrecht, F. T. and Reswick, J. B., Eds., Marcel Dekker, New York, 1977, 173.
43. **Fritsch, G. and Hitzig, E.,** Ueber die elektrische Erregbarkeit des Grosshirns, *Arch. Anat. Physiol. Leipzig*, 37, 300, 1870.
44. **Geddes, L. A., Pearel, J. A., Bourland, J. D., et al.,** Thermal properties of dry metal-foil dispersive electrosurgical electrodes, *J. Clin. Eng.*, 5(1), 13, 1980.
45. **Gilman, S., Dauth, G. W., Tennyson, V. M., and Kremzner, L. T.,** Chronic cerebellar stimulation in the monkey: preliminary observations, *Arch. Neurol.*, 32, 474, 1975.

46. **Gilman, S., Dauth, G. W., Tennyson, V. M., et al.,** Clinical, morphological, biochemical and physiological effects of cerebellar stimulation, in *Functional Electrical Stimulation: Applications in Neural Prostheses,* Hambrecht, F. T. and Reswick, J. B., Eds., Marcel Dekker, New York, 1977, 191.
47. **Glenn, W. W. L., Hageman, J. H., Mauro, A., Eisenberg, L., Flanigan, S., and Harvard, B. M.,** Electrical stimulation of excitable tissue by radiofrequency transmission, *Ann. Surg.*, 160, 338, 1964.
48. **Glenn, W. W. L., Holcomb, W. G., Gee, J. B. L., and Rath, R.,** Central hypoventilation: long-term ventilatory assistance by radiofrequency electrophrenic respiration, *Ann. Surg.*, 172, 755, 1970.
49. **Glenn, W. W. L., Holcomb, W. G., Hogan, J., Matano, I., Gee, J. B. L., Motoyama, E. K., Kim, C. S., Poirier, R. S., and Forbes, G.,** Diaphragm pacing by radiofrequency transmission in the treatment of chronic ventilatory insufficiency, *J. Thorac. Cardiovasc. Surg.*, 66, 505, 1973.
50. **Glenn, W. W. L., Holcomb, W. G., Shaw, R. K., Hogan, J. F., and Holschuh, K.,** Long-term ventilatory support by diaphragm pacing in quadriplegia, *Ann. Surg.*, 183, 566, 1976.
51. **Glenn, W. W. L., Holcomb, W. G., Hogan, J. F., Kaneyuki, T., and Kim, J.,** Long-term stimulation of the phrenic nerve for diaphragm pacing, in *Functional Electrical Stimulation, Applications in Neural Prostheses,* Hambrecht, F. T. and Reswick, J. B., Ed., Marcel Dekker, New York, 1977, 97.
52. **Glenn, W. W. L., Hogan, J. F., and Phelps, M. L.,** Ventilatory support of the quadriplegic patient with respiratory paralysis by diaphragm pacing, *Surg. Clin. N. Am.*, 60(5), 1055, 1980.
53. **Guyton, D. L. and Hambrecht, F. T.,** Capacitor electrode stimulates nerve or muscle without oxidation-reduction reactions, *Science,* 181, 74, 1973.
54. **Grimes, J. H., Nashold, B. S., and Currie, D. P.,** Chronic electrical stimulation of the paraplegic bladder, *J. Urol.*, 109, 242, 1973.
55. **Hochmair-Desoyer, I. J. and Hochmair, E. S.,** Implantable eight-channel stimulator for the deaf, *Proc. ESSCIRC,* 77, 84, 1977.
56. **Hochmair-Desoyer, I. J., Hochmair, E. S., Fischer, R. E., and Burian, K.,** Cochlear prostheses in use: recent speech comprehension results, *Arch. Otorhinolaryngol.*, 229, 91, 1980.
57. **Hochmair-Desoyer, I. and Hochmair, E. S.,** An eight-channel scala tympani electrode for auditory prosthesis, *IEEE Trans. Biomed. Eng.*, 27, 44, 1980.
58. **Horsley, V. and Clark, R. H.,** The structure and functions of the cerebellum examined by a new method, *Brain,* 31, 44, 1908.
59. **Hopkinson, N. T., Bilger, R. C., and Black, F. O.,** Present audiologic status of subjects implanted with auditory prostheses, *Ann. Otol. Rhinol. Laryngol. (Suppl.),* 86(3, Part 3: Suppl. 38), 40, 1977.
60. **House, W. F. and Urban, J.,** Long term results of electrode implantation and electronic stimulation of the cochlea in man, *Ann. Otol. Rhinol. Laryngol.*, 82, 504, 1973.
61. **House, W. F.,** Cochlear implants, *Ann. Otol. Rhinol. Laryngol.*, 85(Suppl. 27), 1, 1976.
62. **House, W. F.,** The clinical value of single electrode systems in auditory prostheses, *Otolaryngol. Clin. N. Am.*, 11(1), 201, 1978.
63. **House, W. F., Berliner, K. I., and Eisenberg, L. S.,** Present status and future direction of the Ear Research Institute cochlear implant program, *Acta Otolaryngol.*, 87, 176, 1979.
64. **House, W. F. and Berliner, K. I.,** Cochlear implants: progress and perspectives, *Ann. Otol. Rhinol. Laryngol.*, 91(2, Part 3: Suppl. 91), 1, 1982.
65. **House, W. F. and Edgerton, B. J.,** A multiple-electrode cochlear implant, *Ann. Otol. Rhinol. Laryngol.*, 91, 104, 1982.
66. **Jako, G. J.,** Electrical stimulation of the human cochlea and the flexible multichannel intracochlear electrode, *Otolaryngol. Clin. N. Am.*, 11(1), 235, 1978.
67. **Jonas, U., Heine, J. P., and Tanagho, E. A.,** Studies on the feasibility of urinary bladder evacuation by direct spinal cord stimulation, *Invest. Urol.*, 13, 142, 1975.
68. **Jonas, U., Jones, L. W., and Tanagho, E. A.,** Spinal cord stimulation versus detrusor stimulation, *Invest. Urol.*, 13, 171, 1975.
69. **Jonas, U., Jones, L. W., and Tanagho, E. A.,** Controlled electrical evacuation via stimulation of the sacral micturition center or direct detrusor stimulation, *Urol. Int.*, 31, 108, 1976.
70. **Kim, J. H., Manuelidis, E. E., Glenn, W. W. L., Fukuda, Y., Cole, D. S., and Hogan, J. F.,** Light and electron microscopic studies of phrenic nerves after long-term electrical stimulation, *J. Neurosurg.*, 58, 84, 1983.
71. **Klomp, G. F., Womack, M. V. B., and Dobelle, W. H.,** Fabrication of large arrays of cortical electrodes for use in man, *J. Biomed. Mater. Res.*, 11, 347, 1977.
72. **Krause, F. and Schaum, H.,** Die spezielle Chirurgie der Gehirnkrankheiten, *Neue Dtsch. Chir.*, 49, 483, 1931.
73. **Krieg, W.,** *Functional Neuroanatomy,* 2nd ed., Blakiston, New York, 1953, 207.
74. **Krieg, W. J. S.,** Prospectus for a stereotaxic and electroneuroprosthetic institute, in *Stereotaxy,* Krieg, W. J. S., Brain Books, Evanston, Ill., 1975, 109.
75. **Krieg, W. J. S.,** Electroneuroprosthesis. History and Forecast, in *Stereotaxy,* Krieg, W. J. S., Brain Books, Evanston, Ill., 1975, 126.

76. **Larson, S., Sances, A., Jr., Cusick, J. F., Meyer, G. A., and Swiontek, T. J.,** Cerebellar implant studies, *IEEE Trans. Biomed. Eng.*, 23, 319, 1976.
77. **Larson, S. J., Sances, A., Jr., Hemmy, D. C., Millar, E. A., and Walsh, P. R.,** Physiological and histological effects of cerebellar stimulation, *Appl. Neurophysiol.*, 40, 160, 1977/78.
78. **Lilly, J. C.,** Injury and excitation by electric currents. A. The balanced pulse-pair waveform, in *Electrical Stimulation of the Brain*, Sheer, D. E., Ed., University of Texas Press, Austin, 1961, 60.
79. **Merzenich, M. M., Michelson, R. P., Pettit, C. R., Schindler, R. A., and Reid, M.,** Neural encoding of sound sensation evoked by electrical stimulation of the acoustic nerve, *Ann. Otol.*, 82, 486, 1973.
80. **Merzenich, M. M., Schindler, D. H., and White, M. W.,** Symposium on cochlear implants. II. Feasibility of multichannel scala tympani stimulation, *Laryngoscope*, 84, 1887, 1974.
81. **Merzenich, M. M.,** Studies on electrical stimulation of the auditory nerve in animals and man: cochlear implants, in *The Nervous System*, Tower, D. B., Ed., Raven Press, New York, 1975.
82. **Merzenich, M. M. and White, M.,** Cochlear implants: the interface problem, in *Functional Electrical Stimulation: Applications in Neural Prostheses*, Marcel Dekker, New York, 1977.
83. **Merzenich, M. M., White, M., Vivion, M. C., Leake-Jones, P. A., and Walsh, S.,** Some considerations of multichannel electrical stimulation of the auditory nerve in the profoundly deaf: interfacing electrode arrays with the auditory nerve array, *Acta Otolaryngol.*, 87(3-4), 196, 1979.
84. **Merzenich, M. M., Byers, C. L., White, M., and Vivion, M. C.,** Cochlear implant prostheses: strategies and progress, *Ann. Biomed. Eng.*, 8, 361, 1980.
85. **Michelson, R. P.,** Electrical stimulation of the human cochlea, *Arch. Otolaryngol.*, 43, 317, 1971.
86. **Michelson, R. P.,** The results of electrical stimulation of the cochlea in human sensory deafness, *Ann. Otol.*, 80, 914, 1971.
87. **Michelson, R. P., Merzenich, M. M., Pettit, C. R., et al.,** A cochlear prosthesis and future clinical observations: preliminary results of physiological studies, *Laryngoscope*, 83, 1116, 1973.
88. **Miller, N. E., Jensen, D. D., and Myers, A.,** Injury and excitation by electric currents. B. A comparison of the Lilly waveform and the sixty-cycle sine wave, in *Electrical Stimulation of the Brain*, Sheer, D. E., Ed., University of Texas Press, Austin, 1961, 64.
89. **Mladejovsky, M. G., Eddington, D. K., Dobelle, W. H., and Brackmann, D. E.,** Artificial hearing for the deaf by cochlear stimulation: pitch modulation and some parametric thresholds, *Trans. Am. Soc. Artif. Intern. Organs*, 21, 1, 1975.
90. **Mortimer, J. T., Shealy, C. N., and Wheeler, C.,** Experimental non-destructive electrical stimulation of brain and spinal cord, *J. Neurosurg.*, 32, 553, 1970.
91. **Mountcastle, V. B., Ed.,** *Medical Physiology*, C. V. Mosby, St. Louis, 1974.
92. **Nashold, B. S., Friedman, H., Glenn, J. F., Grimes, J. H., Barry, W. F., and Avery R.,** Electromicturition in paraplegia, *Arch. Surg.*, 104, 195, 1972.
93. **Nashold, B. S., Grimes, J. H., Friedman, H., and Avery, R.,** Stimulation of the conus medullaris in the paraplegic, *Appl. Neurophysiol.*, 40, 290, 1977/78.
94. **Newman, J.,** Resistance for flow of current to a disk, *J. Electrochem. Soc.*, 113, 501, 1966.
95. **Penfield, W. and Rasmussen, T.,** *The Cerebral Cortex of Man*, Macmillan, New York, 1950.
96. **Penfield, W. and Jasper, J.,** *Epilepsy and the Functional Anatomy of the Human Brain*, Little, Brown, Boston, 1954.
97. **Pialoux, P., Chouard, C. H., and MacLeod, P.,** Physiological and clinical aspects of the rehabilitation of total deafness by implantation of multiple introcochlear electrodes, *Acta Otolaryngol.*, 81, 436, 1976.
98. **Pialoux, P., Chouard, C. H., Meyer, B., and Fugain, C.,** Indications and results of multichannel cochlear implant, *Acta Otolaryngol.*, 87, 185, 1979.
99. **Pudenz, R. H., Bullara, L. A., Dru, D., and Talalla, A.,** Electrical stimulation of the brain. II. Effects on the blood-brain barrier, *Surg. Neurol.*, 4, 265, 1975.
100. **Richardson, R. R., Johnson, N., and Cerullo, L. J.,** Diaphragm pacing in central von Recklinghausen's disease: a case report, *Neurosurgery*, 3, 75, 1978.
101. **Richardson, R. R., Roseman, B., and Singh, N.,** Diaphragm pacing in spinal muscular atrophy: case report, *Neurosurgery*, 9 (3), 317, 1981.
102. **Robertson, L. T., Dow, R. S., Cooper, I. S., and Levy, L. F.,** Morphological changes associated with chronic cerebellar stimulation in the human, *J. Neurosurg.*, 51, 510, 1979.
103. **Rowland, V., MacIntyre, W. J., and Bidder, T. G.,** The production of brain lesions with electrical currents. II. Bidirectional currents, *J. Neurosurg.*, 17, 55, 1960.
104. **Ruch, T. C., Patton, H. D., Woodbury, J. W., and Towe, A. L., Eds.,** *Neurophysiology*, 2nd ed., W. B. Saunders, Philadelphia, 1965.
105. **Sances, A., Jr., Myklebust, J., Larson, S., Darin, J., Swiontek, T., Prieto, T., Chilbert, M., and Cusick, J.,** Experimental electrical injury studies, *J. Trauma*, 21(8), 589, 1981.
106. **Satch, I., Fujii, Y., Kaneyuki, T., Hogan, J. F., Holcomb, W. G., and Glenn, W. W. L.,** Totally implantable diaphragm pacemaker, *Surg. Forum*, 27, 290, 1976.

107. **Schmidt, R. A. and Tanagho, E. A.,** Feasibility of controlled micturition through electric stimulation, *Urol. Int.*, 31, 115, 1979.
108. **Severinghaus, J. W. and Mitchell, R. A.,** Ondine's curse: failure of respiratory center automticity while awake, *Clin. Res.*, 10(Abstr.), 122, 1962.
109. **Shaw, D.,** Methods and Means for Aiding the Blind, U.S. Patent 2, 721, 316, 1955.
110. **Simmons, F. B.,** Electrical stimulation of acoustical nerve and inferior colliculus: results in man, *Arch. Otolaryngol.*, 79, 559, 1964.
111. **Simmons, F. B.,** Electrical stimulation of the auditory nerve in man, *Arch. Otolaryngol.*, 84, 2, 1966.
112. **Susset, J. G., Rottembourg, J. L., Ghoneim, M. A., and Fretin, J.,** Electrical stimulation of the pelvic floor, in *Functional Electrical Stimulation: Applications in Neural Prostheses*, Hambrecht, F. T. and Reswick, J. B., Eds., Marcel Dekker, New York, 1977, 141.
113. **Swiontek, T., Maiman, D., Sances, A., Jr., Myklebust, J., Larson, S., and Hemmy, D.,** Effect of electrical current on temperature and pH in cerebellum and spinal cord, *Surg. Neurol.*, 14(5), 365, 1980.
114. **Tanagho, E. A.,** Induced micturition via intraspinal sacral root stimulation clinical implications, in *Functional Electrical Stimulation: Applications in Neural Prostheses*, Hambrecht, F. T. and Reswick, J. B., Eds., Marcel Dekker, New York, 1977, 157.
115. **Timm. G. and Bradley, W.,** Electrostimulation of the urinary detrusor to effect contraction and evacuation, *Invest. Urol.*, 6, 562, 1969.
116. **Timm, G. and Bradley, W.,** Electromechanical restoration of the micturition reflex, *IEEE Trans. Biomed. Eng.*, 18, 274, 1971.
117. **Tong, Y. C., Black, R. C., Clark, G. M., Forster, I. C., Millar, J. B., O'Loughlin, B. J., and Patrick, J. F.,** A preliminary report on a multiple-channel cochlear implant operation, *J. Laryngol. Otol.*, 93, 679, 1979.
118. **Vanderlinden, R. G., Filpin, L., Harper, J., McClurkin, M., and Twilley, D.,** Electrophrenic respiration in quadriplegia, *Can. Nurse*, 70, 23, 1974.
119. **Weinman, J.,** Biphasic stimulation and electrical properties of metal electrodes, *J. Appl. Physiol.*, 20, 787, 1965.
120. **Weissman, A. D. and Schwartz, E. L.,** A flexible high density multi-channel electrode array for long-term chronic implantation, *Brain Res. Bull.*, 6(6), 543, 1981.
121. **White, R. L.,** Microelectronics and neural prostheses, *Ann. Biomed. Eng.*, 8, 317, 1980.
122. **White, R. L.,** Review of current status of cochlear prostheses, *IEEE Trans. Biomed. Eng.*, 29(4), 233, 1982.
123. **Yuen, T. G. H., Agnew, W. R., Bullara, L. A., Jacques, S., and McCreery, B.,** Histological evaluation of neural damage from electrical stimulation: considerations for the selection of parameters for clinical application, *Neurosurgery*, 9(3), 292, 1981.

Chapter 2

ELECTRICAL FIELD MODELING

Robert Plonsey

TABLE OF CONTENTS

I. BASIC PRINCIPLES

A. Excitable Cells

Excitable cells, such as nerve and muscle, can be described simply as consisting of a uniform conducting intracellular region separated from the extracellular space by a very thin (~100 Å) plasma membrane. The excitable cell properties arise from the unequal intracellular and extracellular ionic composition ([Na]o ≫ [Na]i and [K]o ≪ [K]i) and the selectively permeable membrane properties. At rest the concentration gradients cause ion movements that result in the establishment of a resting potential. Upon stimulation the membrane undergoes a phasic change in permeability resulting in ion movements and the characteristic action potential.[11]

Concomitant to the action potential are action currents, which arise because of the departure of the transmembrane potential, at the stimulus site, from equilibrium. These currents link the intracellular with the extracellular space and their transmembrane direction is such as to depolarize the adjacent membrane resulting in a contiguous conduction of the action potential (propagation). The propagation of impulses is, of course, an essential part of communication from peripheral sensors to the brain and from the brain to organs and to skeletal muscle (and along muscle cells themselves as part of the excitation-contraction process). A quantitative study of propagation requires some way of representing the nonlinear electrical properties of each patch of membrane, the best known being the Hodgkin-Huxley formulation.[6] The perspective in this chapter and book is concerned with the excitatory process, however, in which the cell to be excited is initially passive. In this state, the membrane behavior is linear and the membrane can be modeled as a network consisting of resistor and capacitor elements. For muscle it has been suggested that an appropriate representation is a parallel combination of resistance, capacitance, and a series resistance and capacitance (the latter reflecting the contribution of the transverse tubules). However, for nerve a simple parallel resistance and capacitance is appropriate, and we will assume this in what follows. (In the excitatory process the threshold of nerve is so much lower than muscle that it, alone, need be considered.)

B. The Linear Core Conductor Model

Since only departures from the resting condition are of interest, we can suppress this constant additive value in which case an electrical representation of an excitable unmyelinated axon is that shown in Figure 1. In this figure the subscript i represents intracellular and subscript e represents extracellular conditions. The model rests on the assumption that the intracellular and extracellular currents are confined to the axial direction. One can show this to be well justified intracellularly in view of the long narrow shape of the axoplasm. For fibers in a trunk, the interstitial space is similarly confining and the extracellular approximation is justified. For an isolated axon, extracellular current spreads radially as well as longitudinally and in this case the extracellular representation shown in Figure 1 would not be correct. However, in this case extracellular potentials are small so that, relative to intracellular quantities, one can proceed with setting $r_e = 0$. Application of Kirchoff circuit laws to the linear core conductor model of Figure 1 gives rise to the cable equations below, namely,

$$\partial V_e/\partial x = -I_e r_e \quad (1)$$

$$\partial V_i/\partial x = -I_i r_i \quad (2)$$

$$\partial I_i/\partial x = -i_m \quad (3)$$

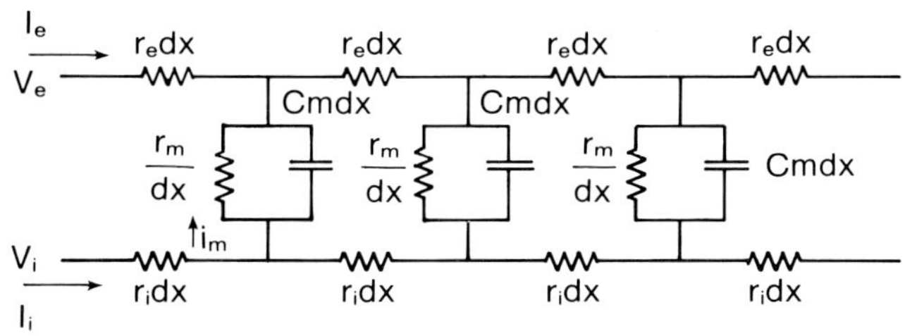

FIGURE 1. The linear core conductor model. Each repetitive element is for a differential axial extent dx. The subscript i denotes intracellular and the subscript e denotes extracellular properties.

where I_i, I_e is the total axial current, i_m the (outward) transmembrane current per unit length, and r_i and r_e are the axial resistance per unit length. The transmembrane potential is conventionally chosen as $V_m = V_i - V_e$ and this leads to

$$\partial^2 V_m/\partial x^2 = (r_i + r_e)i_m + r_e i_p \tag{4}$$

where i_p is the stimulating current per unit length (which we tacitly assume to enter the extracellular space). Assuming subthreshold conditions, then

$$i_m = V_m/r_m + C_m dV_m/dt \tag{5}$$

Substituting Equation 5 into Equation 4 results in

$$\lambda^2 \partial^2 V_m/\partial x^2 - \tau \partial V_m/\partial t - V_m = r_e \lambda^2 i_p \tag{6}$$

where the space constant $\lambda = \sqrt{r_m/(r_i + r_e)}$ and the time constant $\tau = r_m C_m$. For a step current input at a discrete position X_1 and sink at the position X_2 (i.e., $i_p = [\delta(x - x_1) - \delta(x - x_2)]u(t)$, $\delta(x)$ being a unit delta function and $u(t)$ being a unit step function), then the solution to Equation 6 comes out[4]

$$V_m(x,t) = -\frac{1}{2} r_e \lambda [2F(x - x_1,t) - 2F(x - x_2,t)] \tag{7}$$

and

$$F(x,t) = \frac{1}{4}\left\{ e^{-|x|/\lambda}\left[1 - \mathrm{erf}\left(\frac{|x|}{2\lambda}\sqrt{\frac{\tau}{t}} - \sqrt{\frac{t}{\tau}}\right)\right] \right.$$
$$\left. e^{|x|/\lambda}\left[1 - \mathrm{erf}\left(\frac{|x|}{2\lambda}\sqrt{\frac{\tau}{t}} + \sqrt{\frac{t}{\tau}}\right)\right]\right\} \tag{8}$$

In Equation 8, erf is the error function. Note that length and time are normalized by λ and τ, respectively. For $x_1 = -\infty$ and $x_2 = 0$, the spatial and temporal curves reflect the response to a single electrode and show an exponential-like temporal build-up to steady state as well as a steady-state exponential (spatial) decrease in potential with increasing distance from the stimulating electrode.

C. Volume Conductor Fields

A rigorous study of electric (and magnetic) fields in the conducting medium that characterizes physiological preparations starts with Maxwell's equation.[14] Under harmonic conditions and in vector form, these are

$$\nabla x\overline{H} = (\sigma + j\omega\epsilon)\overline{E} + \overline{J}^i \tag{9}$$

$$\nabla x\overline{E} = - j\omega\mu\overline{H} \tag{10}$$

where $\bar{E}$ is the electric field, $\bar{H}$ is the magnetic field, σ is the conductivity, ϵ is the permittivity, μ is the permeability, ω the angular frequency, and $\bar{J}^i$ is the applied (nonconservative) current. The latter is the current introduced by the stimulating electrodes (it may also represent the diffusion-driven ionic current permitted to flow across a membrane whose permeability is temporarily elevated during an action potential, generator potential, etc.). The solution of Equations 9 and 10 is aided by defining auxiliary functions Φ and $\bar{A}$ such that

$$\overline{H} = \nabla \times \overline{A} \tag{11}$$

$$(\overline{E} + j\omega\overline{A}) = -\nabla\Phi \tag{12}$$

Now, for physiological media, the macroscopic permittivity and permeability are nominally that of free space. For frequency components under around 10 kHz, the displacement current is negligible compared to the conduction current ($\omega\epsilon/\sigma \ll 1$), and this is a condition which applies to action, generator, and synaptic potential sources.[16] A consequence is that Equation 12 becomes

$$\overline{E} = -\nabla\Phi \tag{13}$$

and the scalar potential Φ and vector potential $\bar{A}$ satisfy the static (Poisson) equation

$$\nabla^2\Phi = \nabla \cdot \overline{J}^i/\sigma \tag{14}$$

$$\nabla^2\overline{A} = -\overline{J}^i \tag{15}$$

While the fields are time-varying, at each instant the static expressions apply, and the condition is known as quasistatic. In electrostatics if a volume charge distribution ρ lies in a uniform dielectric medium of infinite extent, then

$$\nabla^2\Phi = -\rho/\epsilon \tag{16}$$

and

$$\Phi = \frac{1}{4\pi\epsilon}\int \rho/r \, dV \tag{17}$$

where R is the distance from an element of source, ρdV, to the point at which the field is being evaluated. Now, by duality with the above equations, integral solutions to Equations 14 and 15 (assuming a uniform conducting medium of infinite extent) will be

$$\Phi = \frac{1}{4\pi\sigma}\int -\nabla \cdot \overline{J}^i/r \, dV \tag{18}$$

$$\overline{A} = \frac{1}{4\pi}\int \overline{J}^i/r \, dV \tag{19}$$

In Equation 18, the expression $-\nabla \cdot \bar{J}^i$ can be identified as a current source density, with the same field-generating behavior as ρ in electrostatics. Using vector identities one can also show that

$$\Phi = \frac{1}{4\pi\sigma}\int \overline{J}^i \cdot \nabla(1/r) \, dV \tag{20}$$

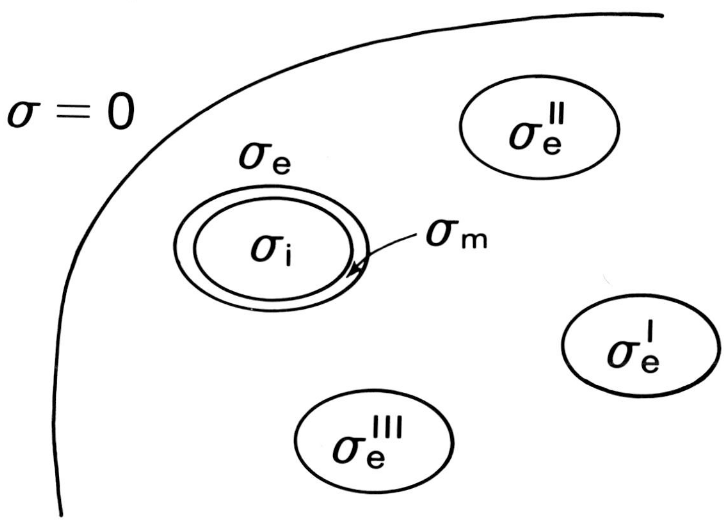

FIGURE 2. The figure reflects physiological inhomogenieties in conductivity on a macroscopic and microscopic basis.

permitting $\bar{J}^i$ also to be interpreted as a volume dipole density distribution.

If we examine Equation 9, we can identify the right-hand side as evaluating the total current $\bar{J}$, namely,

$$\bar{J} = \sigma\bar{E} + j\omega\epsilon\bar{E} + \bar{J}^i \tag{21}$$

and consisting of a conduction current, displacement current, and applied or impressed current. The first two components flow as a consequence of (i.e., secondary to) an existing electric field; the applied current is established by nonelectrical sources and reflects some primary source of energy. Since the divergence of the curl of $\bar{H}$ is necessarily zero, then from Equation 9 it is clear that the total current is divergenceless (solenoidal). Consequently, assuming the displacement current negligible compared to the conduction current,

$$\nabla \cdot \bar{J} = 0 = \sigma\nabla \cdot \bar{E} + \nabla \cdot \bar{J}^i \tag{22}$$

Equation 14 can be derived by substituting Equation 13 into Equation 22.

D. Secondary Sources

Equations 18 and 19 relate sources ($\bar{J}^i$) to the fields they produce (Φ and $\bar{A}$, hence $\bar{E}$ and $\bar{H}$), provided the conductive medium is uniform. Physiological preparations are normally inhomogeneous and this factor must be considered when evaluating the fields set up by the impressed current sources, $\bar{J}^i$. Figure 2 describes the general form of media of interest. The regions σ_e' and σ_e'' suggest the presence of volume elements of different, macroscopic conductivity. A cell is also shown representing the separate intracellular, membrane and extracellular conductivities that arise in this instance. Outside the boundary of the preparation, the conductivity is zero (air). As noted earlier, the permittivity and permeability can usually be assumed uniform, but physiological preparations generally contain the inhomogeneities in conductivity noted here.

The effect of inhomogeneities, such as described in Figure 2, on the electric (and magnetic) field is the imposition of boundary conditions at the interface between media of different conductivity. These boundary conditions correspond to the presence of secondary (induced) sources at the boundary. One way of accounting for the imposed boundary conditions is to add the secondary sources to the original (primary) source, $\bar{J}^i$, in determining the resultant field. The combined primary and secondary sources are then considered to be in a uniform homogeneous region of infinite extent; consequently Equations 19 and 20 apply except that the source is $\bar{J}^i + \Sigma\bar{K}_k^{\,i}$. For the kth, such interface between media of different conductivity σ_k'' and σ_k', the secondary source is a current double layer $\bar{K}_k^{\,i}$ given by[12]

$$\overline{K}_k^i = \Phi_k(\sigma_k'' - \sigma_k')\overline{n} \tag{23}$$

when $\bar{n}$ is the surface normal directed from the primed to double-primed region and Φ_k is the electric potential at the kth surface.

Secondary sources arise not only at the interface between macroscopic regions of different conductivity, but also as suggested in Figure 2, in those which occur on a cellular scale. Thus, at the inner and outer membrane surface of a cell we have, applying Equation 23,

$$\overline{K}_c^i = [\Phi_e(\sigma_e - \sigma_m) - \Phi_i(\sigma_i - \sigma_m)]\overline{n} \tag{24}$$

where Φ_e, Φ_i are the extracellular and intracellular potentials and $\bar{n}$, is the outward normal from the cell. Now since σ_m is very small compared to σ_e or σ_i, Equation 24 reduces to

$$\overline{K}_c^i = (\sigma_e\Phi_e - \sigma_i\Phi_i)\overline{n} \tag{25}$$

In Equation 25 as in Equation 24, the very close separation of secondary source at the outer and inner membrane surface justifies combining both at a location, say, midway between the two. Equation 25 applies for both active as well as passive membranes, but for the latter a constraining condition is Equation 5.

Volume conductor impedance may change as a result of respiration, blood flow, tumor growth, etc. It is useful to follow these changes by applying an external current and measuring the resultant potential field. The changing impedance will be manifest by a changing potential and the potential field due to the changes (alone) can be ascribed to a secondary source which turns out to be given by[15]

$$\overline{J}_p^i = \frac{\Delta\sigma}{\sigma_1\sigma_2}\overline{J}_a \tag{26}$$

In this expression, $\bar{J}_a$ is the applied current density under final conductivity conditions σ_2, while σ_1 is the initial conductivity distribution, and $\Delta\sigma = \sigma_2 - \sigma_1$. Note that $\bar{J}_p^{\,i}$ is a volume source density, in general.

E. Integral Equation Formulation

We consider in this section a mathematical approach to the determination of the induced transmembrane potential of a single isolated cell due to a stimulating field. While cells do not occur in isolation it may, nevertheless, be possible to infer the nature of the stimulating field in the presence of other cells, in which case the approach to be outlined may be used.

We assume the stimulating field to be sinusoidal, hence permitting the complex phasor approach to the membrane impedance. The response to any arbitrary time-varying stimulus can be determined using a Fourier synthesis.

It also turns out that for cells which are spherical-like in shape, or for cells which while cylindrical are not too long, that the intracellular potential is approximately constant (iso-potential). Consequently, in the expression for secondary cellular sources (Equation 25), Φ_i can be neglected and Φ_e can be set equal to $-V_m$ (since $V_m = \Phi_i - \Phi_e \simeq -\Phi_e$).

In Figure 3, an arbitrary shaped cell is shown, and we assume it lies in an applied (stimulating) potential field Φ_a. The potential at an arbitrary external field point p′ is then

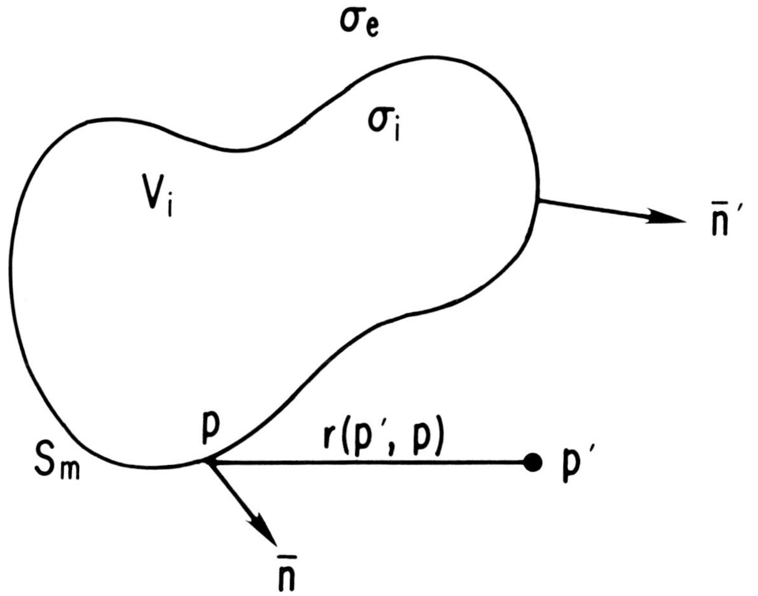

FIGURE 3. Cell geometry and nomenclature. Field points are primed (p′) and those which lie on the membrane surface (S_m) are associated with primed unit outward normals ($\bar{n}'$); source points are unprimed (p) and are associated with unprimed normals ($\bar{n}$); r (p′,p) is the distance from the field point (p′) to the source point (p); σ_i is the conductivity of the intracellular and σ_e is the extracellular media (reciprocal ohms per centimeter); V_i is the intracellular volume. (From Klee, M. and Plonsey, R., *Biophys. J.*, 12, 1676, 1972. With permission.)

equal to the sum of the applied potential field plus that due to the secondary double layer specified by Equation 25 (with Φ_e replaced by $-V_m$ and $\Phi_i = 0$) utilizing Equation 20. The result can be put into the following form,[8]

$$\Phi(p') = \Phi_a(p') - \frac{1}{4\pi}\int_{S_m} V_m \frac{\partial[1/r(p',p)]}{\partial n}\, dS \tag{27}$$

where r(p′,p) is the distance from the field point to integration (source) point and S_m is the cell membrane surface. The normal current density at the membrane surface is found by differentiating Equation 27 with respect to the normal n′:

$$J_m(n') = -\sigma_e \partial\Phi_a(p')/\partial n' + \frac{\sigma_e}{4\pi}\frac{\partial}{\partial n'}\int_{S_m} V_m \frac{\partial[1/r(p',p)]}{\partial n}\, dS \tag{28}$$

The normal current density can also be written by application of Ohm's law,

$$J_m(n') = V_m(\sigma_m + j\omega C_m) \tag{29}$$

where σ_m is the specific membrane conductance (mhos/cm^2) and C_m the membrane capacitance (farads per square centimeter). Equating Equations 28 and 29 yields a Friedholm integral equation for the desired transmembrane potential.

To solve the aforementioned integral equation, one differentiates under the integral sign in Equation 28, interchanges the order of differentiation, discretizes the membrane surface into N elements, and approximates the integral by a finite sum over the N elements. The result is N equations in the N unknown samples of transmembrane potential each of the form

$$(\sigma_m + j\omega C_m)V_m(j) = -\sigma_e\partial\Phi_a/\partial n_j + \frac{\sigma_e}{4\pi}\sum_{k=1}^{n} V_m(k)\frac{\partial^2(1/r_{jk})}{\partial n_j\partial n_k}\Delta S_k \tag{30}$$

where the source point is designated by k and the field point is designated by j. A study of some of the properties of this equation and its application to a spherical cell in an applied uniform field is found in Klee and Plonsey[8] and also discussed in a subsequent section.

While Equation 30 was derived on the assumption, frequently satisfied, of an isopotential intracellular region, it is also generally valid when intracellular and extracellular conductivity are equal. For a circular cylindrical cell, we discuss in the next section how the potential field can be calculated from V_m regardless of σ_i and σ_e. In the most general case, the problem can be approached by iteration.[12]

When the cell is essentially isopotential, then this implies little transmembrane current so that the cell behavior, as seen from the outside, is essentially that of an insulator. In this case, the left-hand side of Equation of 30 can be set equal to zero and we have

$$0 = -\sigma_e\partial\Phi_a/\partial n_j + \frac{\sigma_e}{4\pi}\Sigma\Phi_e(k)\frac{\partial^2[1/r(p',p)]}{\partial n_j\partial n_k}\Delta S_k \tag{31}$$

From the solution for (in this case) Φ_e, one can find Φ_i (hence V_m) though the requirement that the total current crossing the membrane be zero. This requires that

$$\int_{S_m}(\sigma_m + j\omega C_m)(\Phi_i - \Phi_e)\,dS = 0 \tag{32}$$

and since we assume Φ_i to be isopotential, we are led to

$$\Phi_i = (1/A_{cell})\int \Phi_e\,dS \tag{33}$$

where A_{cell} is the surface area of the cell, and σ_m, C_m are assumed to be uniform over the cell surface.

II. FIELDS GENERATED BY ACTIVE FIBERS

A. Intracellular/Extracellular Relationships

When stimulated, a nerve axon or muscle fiber will carry a propagating action potential. Associated with the action potential are action currents which flow throughout the volume conductor. The action currents reflect the presence of an electrical potential field which also pervades the volume conductor. Since noninvasive potential measurements may be made at the surface of the body, it is of interest to develop a theoretical relationship between such potentials and the underlying action potential sources. This forms the basis for physiological and/or clinical interpretation of such surface potential recording. We consider, in this section, the field of a single circular cylindrical fiber of infinite length with the recognition that this solution can form the basis for treatment of more complex fiber structures.

The problem at hand is to investigate the intracellular/extracellular field relations associated with a propagating action potential on a uniform circular cylindrical fiber of radius a, lying in a uniform homogeneous conducting medium. If Equation 20 is used with the secondary source described by Equation 25, then the result is

$$\Phi(P) = \frac{1}{4\pi\sigma_e}\int(\sigma_e\Phi_e - \sigma_i\Phi_i)\nabla(1/R)\cdot d\bar{S} \tag{34}$$

where the surface integration is over the membrane surface, site of the secondary current source, and P is the extracellular field point. For the isolated axon in particular, $\Phi_e \ll \Phi_i$ and Equation 34 simplifies to

$$\Phi(P) = \frac{-\sigma_i}{4\pi\sigma_e} \int \Phi_i \nabla(1/R) \cdot d\bar{S} \tag{35}$$

Under the reasonable condition that Φ_i varies with x only, Equation 35 can be transformed into two alternate forms. These are[12]

$$\Phi(P) = -\frac{1}{4\pi}\frac{\sigma_i}{\sigma_e} \int_A dA \int_{-\infty}^{\infty} \frac{\partial \Phi_i}{\partial x} \bar{a}_x \cdot \nabla(1/R)dx \tag{36}$$

and

$$\Phi(P) = \frac{1}{4\pi}\frac{\sigma_i}{\sigma_e} \int_A dA \int_{-\infty}^{\infty} \frac{\partial^2 \Phi_i}{\partial x^2}\frac{1}{R}\, dx \tag{37}$$

These integrations are through the cell volume and have been separated into a cross-sectional (dA) and axial (dx) part. In Equation 36, the field is seen to arise from an equivalent source consisting of stacks of double-layer disks oriented in the positive x direction, each with a thickness dx and with a dipole volume source density of $-\partial\Phi_i/\partial x$. On the other hand, the source associated with Equation 37 corresponds to a stack of single-layer source disks, where the source density function is given by $\partial^2\Phi_i/\partial x^2$.

In Figure 4A, Φ_i is described by a monophasic action potential; its first and second derivatives as well as the corresponding equivalent sources, as noted above, are also shown. When the action potential can be well approximated by a triangular shape, then the resultant equivalent single layer sources are discrete disks located at the triangle apexes, as illustrated in Figure 4B. The relative magnitude of each positive disk source is the slope of the correponding triangular segment; for a scalene triangle, the source and hence field will be asymmetrical. The negative disk magnitude equals the sum of the positive disks; the total field is consequently asymptotically quadrupolar.[12]

In Equation 37, the function (1/R) behaves like an approximate delta function when the field point is at the fiber surface, since R can be very small. The potential at the fiber surface being a convolution of $\partial^2\Phi_i/\partial x^2$ with this approximate delta function can be expected to resemble $\partial^2\Phi_i/\partial x^2$ expect being more or less smoothed, and this corresponds to observation. For a monophasic action potential, the surface potential is consequently triphasic corresponding to $\partial^2\Phi_i/\partial x^2$, as shown in Figure 4.

If the core conductor model is used to evaluate transmembrane potential, and under the condition $\Phi_i \simeq V_m$ ($\Phi_e \simeq 0$), then taking the derivative of Equation 2 and substituting into Equation 3 gives

$$i_m = (1/r_i)\partial^2\Phi_i/\partial x^2 \tag{38}$$

where i_m is the transmembrane current per unit fiber length. If the field point is at least several fiber radii distant, the cross-sectional integration in Equation 37 can be approximated by the area, πa^2, alone. Since $r_i = 1/(\sigma_i \pi a^2)$, then Equation 38 reduces Equation 37 to

$$\Phi(P) = \frac{1}{4\pi\sigma_e} \int_{\infty}^{\infty} (i_m/R)\, dx \tag{39}$$

B. Fourier Transform Methods (Circular Fiber)

For the circular cylindrical fiber of infinite extent and assuming axial symmetry, an alternate approach to the study of intracellular/extracellular fields is available based on solutions to Laplace's equation in cylindrical coordinates. We start with Equation 14, noting that $\bar{J}^i = 0$ except in the membrane so that in both intracellular and extracellular space

$$\nabla^2\Phi(\rho,x) = 0 \tag{40}$$

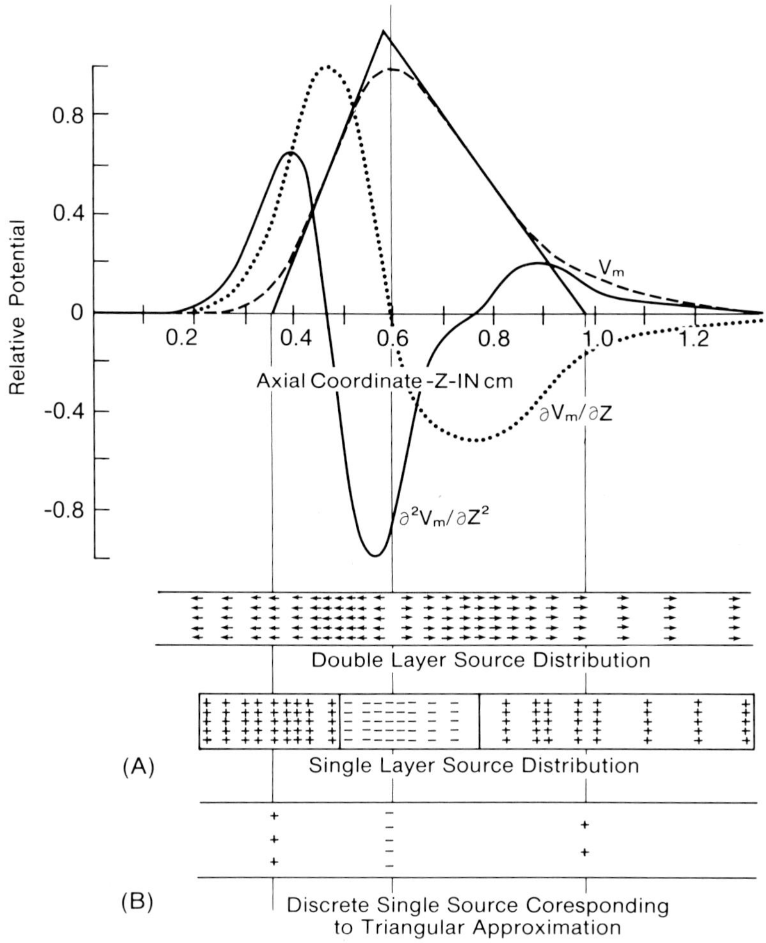

FIGURE 4. (A) The monophasic action potential V_m and its first two derivatives are shown. The equivalent source dipole moment density is proportional to the first derivative, while the equivalent single source density is proportional to the second derivative. The approximating triangular action potential is also shown. In (B) the source corresponding to the second derivative of the triangular approximation is shown, which consists of discrete disk sources of polarity indicated. The positive strengths are proportional to the triangle slope, while the negative strength equals the sum of the positive strengths. This is the "tripolar model".

Using the standard method of separation of variables (Sturm-Liouville theory) one obtains for the extracellular potential field:[2]

$$\Phi_e(\rho,x) = \int_{\infty}^{\infty} B(k)K_o(|k|\rho)e^{-jkx}\,dk \tag{41}$$

and for the intracellular field:

$$\Phi_i(\rho,x) = \int_{\infty}^{\infty} A(k)I_o(|k|\rho)e^{-jkx}\,dk \tag{42}$$

In these equations, K_o is the modified Bessel function of the second kind and I_o is the modified Bessel function of the first kind, ρ is the cylindrical radial variable, and k is the separation constant (eigenvalue) which in this infinite fiber case is continuous.

Now at $\rho = a$, the radial (transmembrane) current must be continuous, i.e.,

$$\sigma_i \partial\Phi_i/\partial\rho \Big|_{\rho=a} = \sigma_e \partial\Phi_e/\partial\rho \Big|_{\rho=a} \tag{43}$$

hence

$$\sigma_i \int_{-\infty}^{\infty} |k| A(k) I_1(|k|a) e^{-jkx} dk = -\sigma_e \int_{-\infty}^{\infty} |k| B(k) K_1(|k|a) e^{-jkx}\, dk \tag{44}$$

This requires that

$$A(k) = \frac{-B(k)K_1(|k|a)\,\sigma_e}{I_1(|k|a)\,\sigma_i} \tag{45}$$

We also require the following to be true at $\rho = a$, namely,

$$V_m(x) = \Phi_i(a,x) - \Phi_e(a,x)$$

$$= \int_{-\infty}^{\infty} [A(k)I_o(|k|a) - B(k)K_o(|k|a)]e^{-jkx}\, dk \tag{46}$$

In Equation 46, $V_m(x)$ can be thought of as the Fourier transform of the bracketed expression. Accordingly the inverse transform yields

$$2\pi[A(k)I_o(|k|a)] - B(k)K_o(|k|a)] = V(k) = \int_{-\infty}^{\infty} V_m(x)e^{jkx}\, dx \tag{47}$$

Equations 45 and 47 can be solved for A(k) and B(k). The final result for Φ_e (ρ,x) is[5]

$$\Phi_e(\rho,x) = \frac{1}{2\pi}\int_{-\infty}^{\infty} \frac{V(k)K_o(|k|\rho)e^{-jkx}dk}{\alpha|k|a)K_o(|k|a)} \tag{48}$$

where

$$\alpha(|k|a) = -\left[1 + \frac{\sigma_e K_1(|k|a)I_o(|k|a)}{\sigma_i K_o(|k|a)I_1(|k|a)}\right] \tag{49}$$

Equation 48 is a formal solution for the extracellular potential given (only) the transmembrane potential. In contrast Equation 34 requires a separate specification of Φ_e and Φ_i.

C. Inverse Problem (Circular Fiber)

In the usual situation the extracellular potential (surface potential) is measured and used to deduce the nature of the sources responsible. An example pertinent here is the measurement of the extracellular potential at some radius R, from which the transmembrane potential is to be sought. Under ideal conditions we have from Equation 41,

$$\Phi_e(R,x) = \int_{-\infty}^{\infty} B(k)K_o(|k|R)e^{-jkx}\, dk \tag{50}$$

Taking the inverse transform of both sides of Equation 50 yields

$$B(k)K_o(|k|R) = \frac{1}{2\pi}\int_{-\infty}^{\infty} \Phi_e(R,x)e^{jkx}\, dx = F_e(kR) \tag{51}$$

$$\Phi_e(\rho,z) = \int_{-\infty}^{\infty} \frac{F_e(kR)}{K_o(|k|R)} K_o(|k|\rho)e^{-jkx}\, dk \tag{52}$$

Equation 52 permits a determination of the entire extracellular potential pattern from the single measurement of Φ_e (R,x).

Comparison of Equation 52 with Equation 48 shows that $F_e(kR)/K_o(|k|R) = V(k)/[\alpha(|k|a)K_o(|k|a)]$, so that

$$V(k) = \frac{\alpha(|k|a)K_o(|k|a)F_e(kR)}{K_o(|k|R)} \tag{53}$$

The transmembrane potential is recovered by taking the inverse transform of V(k)

$$V_m(x) = \frac{1}{2\pi}\int_{-\infty}^{\infty} V(k)e^{jkx}\, dk \tag{54}$$

In Equation 53 the potential at $\rho = R$, represented by its Fourier transform $F_e(kR)$, arises from the transmembrane potential source as input. Accordingly the "system function" $H(|k|R)$ is obtained from

$$H(|k|R) = \frac{K_o(|k|R)}{\alpha(|k|a)K_o(|k|a)} \tag{55}$$

In the idealized inverse process, the reciprocal of $H(|k|R)$ is multiplied by $F_e(kR)$ to obtain V(k). However, real measurements are always contaminated with noise. Assuming it to be additive, then the measured potential $\Phi_m(R,x)$ is

$$\Phi_m(R,x) = \Phi_e(R,x) + n(x) \tag{56}$$

The optimum inverse filter is no longer $1/H(|k|R)$, but a function G(kR) obtained from Wiener-Hopf least squares estimation theory, where[1]

$$G(kR) = \frac{S_{mm}(k)H(|k|R)}{|H(|k|R)|^2 S_{mm}(k) + S_{nn}(k)} \tag{57}$$

and S_{mm} (k) is the spectral density of the transmembrane potential and S_{nn} (k) is the spectral density of the noise (assumed to be white noise and statistically independent of the signal). An example of the application of this inverse process is given in Figure 5.

If the transmembrane potential is everywhere constant, then the extracellular potential is zero as can be seen from Equations 36 and 37 or 47 and 48. Since the variable k corresponds to the angular spatial frequency, a DC transmembrane potential is the condition for $k = 0$. Since $H(|k|R)$ is necessarily zero for $k = 0$, then $G(kR) = 0$, and it is impossible to recover the DC resting potential with the optimum inverse filter. This result is obvious on physical grounds; the DC resting potential can only be measured with an intracellular electrode.

III. TRANSMEMBRANE POTENTIAL INDUCED IN NERVE AND MUSCLE

A. Introduction

In this concluding section, the basic principles presented and developed in the preceding sections will be applied to several simple and stylized examples of nerve and muscle stimulation. While the models to be discussed are very simple, they provide some useful generalizations both as to approach as well as in their conclusions. The availability of the computer along with good theoretical structure makes feasible the simulation of many problems of interest; few appear to have been tackled to date.

In what follows, the passive cell membrane will be sometimes taken to be resistive and its capacitance ignored. For steady-state solutions, there is no approximation, of course. Under some conditions, the capacitance may, indeed, have a considerable effect as will be seen in the first example. However, in the general situation, any solution involving a membrane resistance can be converted, formally, to one involving membrane impedance at some angular frequency ω. The use of a Fourier integral permits the solution to be extended to an arbitrary stimulus waveform.

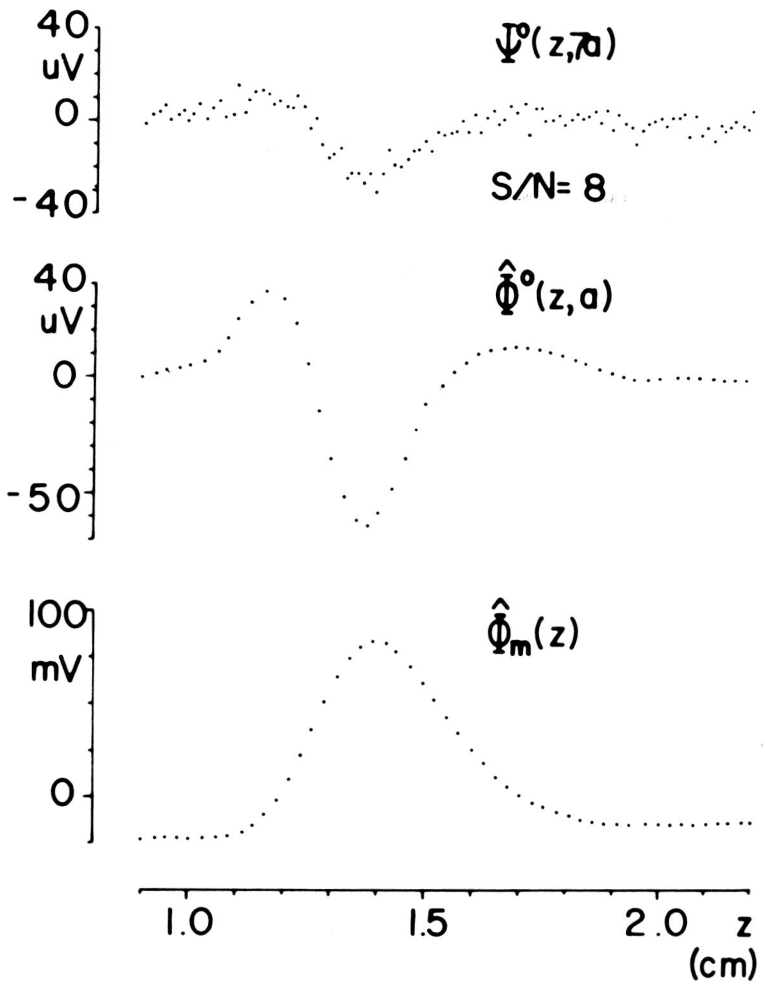

FIGURE 5. Inverse solution given the potential distribution at a radius 7a plus additive white Gaussian noise (S/N = 8) The surface potential Φ^O and the transmembrane potential Φ_m are obtained using Wiener filtering. (From Greco, E. C., Clark, J. W., and Harman, T. L., *Math. Biosci.*, 33, 235, 1977. With permission.)

B. Interaction Among Fibers in Bundle

We consider a large bundle of similar fibers and assume that all but one are carrying an action potential. We are interested in determining the potential induced in the inactive fiber from the remaining active ones. The action potential is assumed to be synchronous on all fibers so that if there are N fibers, then each fiber is associated with 1/N of the interstitial space. Since, normally, this area is comparable to intracellular space, both intracellular and extracellular (interstitial) current flow will be axial and the linear core conductor model (Equations 1 to 4) is well justified. In particular, if the total interstitial area is A_e, then

$$r_e = N/\sigma_e A_e \tag{58}$$

is the extracellular resistance per unit length per fiber. The core conductor equations can be shown to lead to the following interstitial expression:[7]

$$\Phi_e(x) = -\frac{r_e}{r_e + r_i} V_m(x) \tag{59}$$

where r_i is the intracellular resistance per unit length of a typical fiber.

The inactive fiber can also be described by a linear core-conductor model except that its interstitial potential must be given by Equation 59, while its (passive) membrane is a parallel combination of R_m and C_m. A solution is available through a rigorous treatment, but an approximate, intuitive approach is also possible and we will follow this here.[3]

We assume first that the transmembrane potential on the inactive fiber is small. This implies that its intracellular potential must approximate the extracellular applied potential given by Equation 59. Cable equations (i.e., Equation 1) give the intracellular axial current as

$$i_i = -\pi a^2 \sigma_i \partial\Phi_e/\partial x \tag{60}$$

where a is the fiber radius and σ_i is the intracellular conductivity. The transmembrane current per unit length, from Equation 3, is then

$$i_m = -\partial i_i/\partial x = \pi a^2 \sigma_i \partial^2\Phi_e/\partial x^2 \tag{61}$$

On a per unit area basis, the transmembrane current, J_m, is

$$J_m = i_m/2\pi a = (a\sigma_i/2)\partial^2\Phi_e/\partial x^2 \tag{62}$$

If we assume that this transmembrane current is carried mainly by the membrane capacitance, C_m, then the induced transmembrane potential, V_{mi}, is

$$V_{mi}(x) = \int \frac{J_m}{C_m}\, dt = \int \frac{J_m}{C_m}\frac{dx}{\theta} \tag{63}$$

where θ is the assumed uniform velocity of propagation. Substituting Equation 62 into Equation 63 and performing the integration results in

$$V_{mi} = \frac{a\sigma_i}{2C_m}\frac{\partial\Phi_e}{\partial x} \tag{64}$$

Some insight into the order of magnitude of quantities can be obtained through a synthetic example. We assume a fiber of radius 10 μm, $C_m = 0.8\ \mu f/cm^2$, $\theta = 10$ m/sec, $\sigma_i = 1/90(\Omega\text{-cm})^{-1}$, and $r_e/(r_e + r_i) = 0.1$. If we choose the action potential duration to be 1 msec (hence the spatial extent of the action potential is around 1 cm), then a simple and useful expression for Φ_e is

$$\Phi_e(x) = -10e^{-16x^2}\ \text{mV} \tag{65}$$

Substituting these parameters into Equation 62 shows V_{mi} to be at most 0.24 mV, which is relatively small compared to the 10 mV in the interstitial space (as was initially assumed).

For a membrane resistance of $R_m \Omega\text{-cm}^2$, were this the sole pathway of current, we should then replace Equation 63 by $J_m R_m$ or $R_m(a\sigma_i/2)\partial^2\Phi_e/\partial x^2$. Choosing $R_m = 2000\ \Omega\text{-cm}^2$ and other parameters as given above leads to $V_{mi} = 1.6$ mV. This confirms the assumption that the capacitive to resistive current is large ($\sim$7) and justifies use of the capacitance alone. That, indeed, $V_{mi} \propto a/C_m$ can be verified by a more rigorous treatment.[3]

C. Stimulation of Spherical and Spheroidal Cells

For a spherical cell in a stimulating field that can be considered to be uniform, an analytic closed form solution can be obtained. While this is a highly idealized problem, nevertheless, it does provide some insight to more realistic problems through the relationships that exist among the parameters of size, conductivities, and membrane admittance.

For definiteness we choose the polar axis in the direction of the applied electric field and call this the z axis. Then the applied field, Φ_a, can be written

$$\Phi_a = E_o z = E_o r\cos\theta \tag{66}$$

where r, θ are spherical coordinates. The secondary field set up by the sphere must be axially symmetric, as well as a solution of Laplace's equation, hence, must have the form $P_n(\cos\theta)/r^{n+1}$ outside the sphere and $r^n P_n(\cos\theta)$ inside. Since the applied field has a simple cosθ variation, it can only elicit a similar θ variation in the secondary fields, consequently, n = 1 only. The total field, consequently, for a spherical cell of radius a is[14]

$$\Phi_e = E_o r\cos\theta + (A/r^2)\cos\theta \qquad r > a \tag{67}$$

$$\Phi_i = B\cos\theta \qquad r < a \tag{68}$$

where A and B are constants to be determined (B contains E_o plus an undetermined, as yet, secondary contribution).

The constants can be found from the membrane boundary conditions, namely, the continuity of the normal component of current

$$-\sigma_i \partial\Phi_i/\partial r \Big|_{r=a} = -\sigma_e \partial\Phi_e/\partial r \Big|_{r=a} \tag{69}$$

and the membrane admittance relationship (where Y is the admittance in mhos/cm²)

$$\left(\Phi_i \Big|_{r=a} - \phi_e \Big|_{r=a}\right)Y = -\sigma_i \partial\Phi_i/\partial r \Big|_{r=a} \tag{70}$$

If Equations 67 and 68 are substituted into Equations 69 and 70, two linear independant equations in the unknown A, B result. If this is solved, the result is

$$\Phi_i = E_o r\cos\theta \left[1 - \frac{2\sigma_i + RY(\sigma_i/\sigma_e - 1)}{2\sigma_i + RY(\sigma_i/\sigma_e + 2)}\right] \tag{71}$$

$$\Phi_e = E_o r\cos\theta \left[2 - \frac{a^3}{r^3} \frac{1}{\dfrac{2\sigma_e + \sigma_i}{3\sigma_i} + \dfrac{2\sigma_e}{3aY}}\right] \tag{72}$$

Some insight into the significance of these terms follows from substituting typical numerical values ($R_m = 1000\ \Omega\ \text{cm}^2$, a = 15 μm, $\sigma_i = 0.01$ mhos/cm, $Y = 1/R_m$) into Equations 71 and 72. We find that $aY/\sigma_i \ll 1$ and that, assuming $\sigma_e = O(\sigma_i)$,

$$\Phi_i \simeq E_o r\cos\theta \cdot (3aY/2\sigma_i) = E_o r\cos\theta\,(2.25 \times 10^{-4}) \tag{73}$$

$$\Phi_e \simeq E_o r\cos\theta \cdot (1 + a^3/2r^3) \tag{74}$$

The cell interior has a very small field (relative to that applied) and is essentially isopotential (as noted in an earlier section), consequently the cell behaves as if it were an insulated sphere. The extracellular potential at the cell surface is augmented by 50% over its unperturbed (applied) value; for field points at 3 radii or greater, the cell hardly affects the applied field. The induced transmembrane potential is simply

$$\Phi_m = \Phi_i(a,\theta) - \Phi_e(a,\theta) = -1.5\ E_o a\cos\theta \tag{75}$$

The quantity $\lambda_s = \sigma_i/Y$ plays the role of space constant. For the above parameters, $\lambda_s = 100$ cm and consequently a spherical cell with radius small in comparison will be isopotential. For a comparable cylindrical cell of radius a, we found (see Section I.B) that $\lambda_c = \sqrt{\sigma_i a/2Y_m}$ and for a, even as large as 200 μm, $\lambda_c = 1$ cm, showing the importance of cell shape in an examination of the isopotentiality condition.

The induced transmembrane potential on a spherical cell in an applied field can also be found using the integral equation method described in Section I.E. There is obviously no advantage in this approach for this particular cell shape except that it permits an examination of the integral equation method — which works equally well for cells of arbitrary shape. Such was done by Klee and Plonsey[8] and they point out the existence of two singularities that must be dealt with in applying the integral equation approach.

One singularity is associated with the transfer matrix and arises when $Y/\sigma \rightarrow 0$. The difficulty, from a physical standpoint, arises because the solution is unchanged by addition of a constant to the transmembrane potential. This can be overcome through the imposition of the constraint that the total transmembrane current be zero (since the total current must be solenoidal).

The second singularity involves the correct evaluation of the secondary potential field from a patch of membrane at a field point which also lies at that patch (this is denoted the "self-term"). Since a uniform closed double layer generates zero external field, it turns out that the preferable way to evaluate the self-term is to take the negative of the contributions of all remaining membrane elements that form the closed cell surface.

The response of a spherical cell having a nonuniform membrane resistance, consisting of a low-resistance patch in an otherwise high resistance membrane, to a uniform stimulating field, has also been examined.[9] The presence of the patch was shown to cause a shift in the isopotential value of the intracellular region toward the value of the potential just outside the patch. The quantitative effect depends upon the relative change in conductivity of the patch, the area of the patch, the area of the cell, and the extracellular potential distribution. In general the effect is to enhance the transmembrane potential at a membrane site diametrically opposite the low-resistance patch.

A study of the transmembrane potentials induced on spheroidal cells was also made by Klee and Plonsey.[10] In Figure 6, the transmembrane potential as a function of polar angle for a uniform axially applied field is shown for prolate, oblate, and spherical cells. In this example the dimension of each spheroid along its polar axis was the same. The importance of the cell shape is clear from this plot. Some conclusions reached from this study were

1. The less streamlined oblate spheroid generates extracellular potentials which are greater than that of the sphere or prolate spheroid.
2. For each cell type, the extracellular potentials are roughly linearly proportional to cell size.
3. So long as $\lambda = [\sigma_i \text{Vcell}/Y\ \text{Acell}]^{1/2}$ is large, where Vcell is the cell volume and Acell is the surface area, the induced transmembrane potential is independent of the electrical parameters, and the cell interior will be isopotential.

D. Stimulation of a Fiber Bundle

Figure 7A illustrates the use of a cuff electrode to stimulate fibers in a nerve trunk. While discussion of this example is of interest itself, the example also provides some insight into the general behavior of axial current flow which divides between interstitial and intracellular space.

In Figure 7B we show an approximate equivalent electrical system to that in Figure 7A (we will discuss this approximation later) which assumes, essentially, that the interstitial potentials established by each extracellular electrode are uniform across the bundle. With

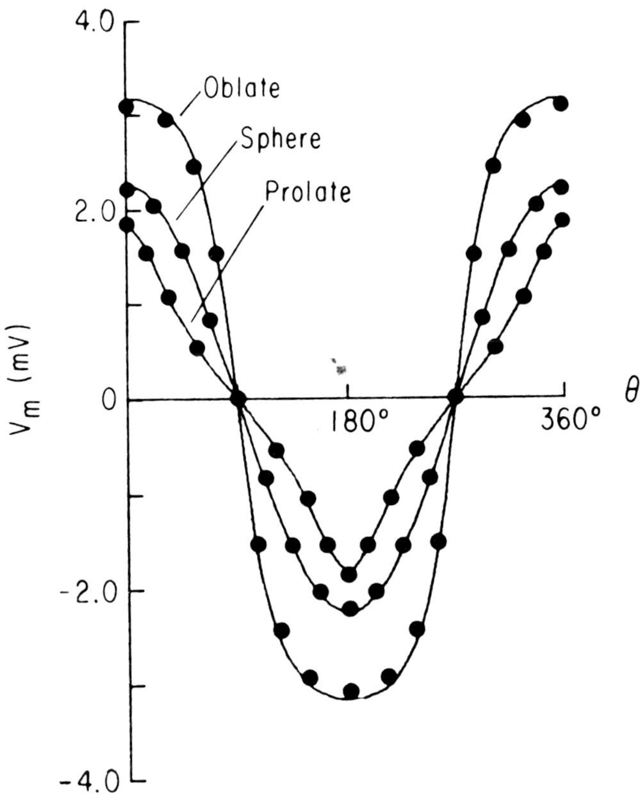

FIGURE 6. Induced transmembrane potential as a function of polar angle. Parameters are for a cell height (in direction of applied field) equal to 30 μm, applied field of 1.0 V/m, σ = 0.05 mhos/cm, σ_m = Y = 0.0005 mhos/cm^2. Data points represent numerical results; solid curve represents an approximate solution. (From Klee, M. and Plonsey, R., *Trans. IEEE Biomed. Eng.*, 23, 347, 1976. With permission.)

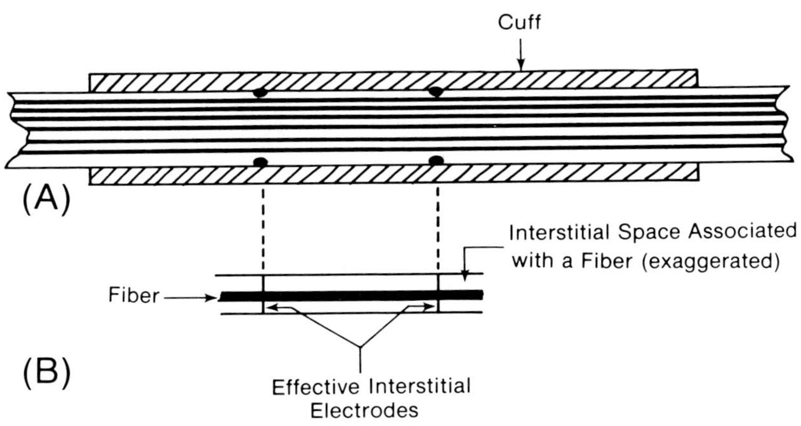

FIGURE 7. (A) Use of cuff/ring electrode for stimulation and block of a fiber bundle. Cuff is assumed to be long and fibers uniform so that the behavior of each fiber is similar. As a result it is only necessary to consider a typical fiber and its associated interstitial space as shown in (B).

this assumption the extracellular (interstitial) and intracellular potentials can be assumed functions of the axial variable, x, alone (at least under the cuff which we consider, for simplicity, to be very long). In fact it is possible, assuming the existence of a large number of essentially identical fibers, to consider a single typical fiber and its associated share of the interstitial space. Since the intracellular and interstitial currents are entirely axial, the linear core conductor model of Section I.B applies.

Under steady-state conditions, a single electrode contributes to each fiber a steady current I_o and if the width of the electrode is very small, the current density function can be expressed as a delta function. (For N identical fibers, I_o is the total stimulating current divided by N.) For the electrode at the origin, the aforementioned current density, i_p, is

$$i_p = I_o\delta(x) \tag{76}$$

Under steady-state conditions ($\partial/\partial t = 0$), the resulting differential equation is Equation 6 subject to Equation 76 and is

$$\lambda^2 d^2V_m/dx^2 - V_m = r_e\lambda^2 I_o\delta(x) \tag{77}$$

where, it will be recalled, $\lambda^2 = r_m/(r_i + r_e)$. The homogeneous solution to Equation 77 is simply

$$V_m = Ae^{-|x|/\lambda} \tag{78}$$

The constant, A, in Equation 78 can be found by integrating Equation 77 from a small distance to the left of the origin to a small distance to the right. For this situation, since V_m is continuous, it gives no contribution. The remaining terms give

$$dV_m/dx \Big|_{x=0^+} - dV_m/dx \Big|_{x=0^-} = r_e I_o \tag{79}$$

Substituting Equation 78 into Equation 79 provides the necessary relationship that

$$\frac{2A}{\lambda} = r_e I_o \tag{80}$$

and we finally have

$$V_m = \frac{r_e\lambda I_o e^{-|x|/\lambda}}{2} \tag{81}$$

For the two electrodes with the anode at $x = -D$ and cathode at $x = +D$, then by superposition of Equation 81 we have

$$V_m = \frac{r_e\lambda I_o}{2} |e^{-|D-x|/\lambda} - e^{|D-x|/\lambda}| \tag{82}$$

Under our assumed passive, subthreshold condition, the transmembrane current depends on V_m and the membrane resistance r_m through

$$i_m = V_m/r_m \tag{83}$$

The intracellular axial current can be found by integrating Equation 3, the result being

$$I_i = \frac{r_e\lambda^2 I_o}{2r_m} [e^{-|D-x|/\lambda} - e^{|D-x|/\lambda}] \tag{84}$$

Based on Equations 82 to 84, a quantitative picture of the stimulating process can be generated. In particular one notes an inward transmembrane current under the anode and consequent hyperpolarization, while under the cathode the reverse takes place. The maximum depolarization that is produced depends on the electrode separation when $D < \lambda$, but if $D \gg \lambda$, then it reaches the maximum value of $r_e\lambda I_o/2$. Also, when $D \gg \lambda$, there is a large region between the electrodes where $V_m = i_m = 0$ and the axial currents divide between intracellular and extracellular space in inverse proportion to their per unit length resistance. It is possible to design the electrode system so that it will excite under the cathode and block under the anode for unidirectional stimulation, but the effect of a realistic length cuff is also important in this analysis (which has been ignored here).[17]

The assumption that the entire interstitial plane under the ring electrode is isopotential can be examined by considering a cuff of infinite extent with a uniform conducting medium (rather than a fiber bundle) inside. Such a problem has a straightforward analytic solution found by using Equation 42 along with the boundary condition Equation 76, which translates into

$$\sigma_i \partial\Phi_i/\partial\rho \Big|_{\rho=a} = \sigma_i \int_{-\infty}^{\infty} |k|A(k)I_1(|k|a)e^{-jkx}\,dk = I_o\delta(x) \tag{85}$$

where a is the radius of the cuff. Taking the Fourier transform shows that

$$A(k) = I_o/(\sigma_i|k|I_1(|k|a)) \tag{86}$$

Consequently

$$\Phi(\rho,x) = 2(I_o/\sigma_i)\int_0^{\infty} \frac{I_o(k\rho)}{kI_1(ka)}\cos(kx)dk \tag{87}$$

where we have taken advantage of the symmetry.

An examination of the potential field of Equation 87 can be made using numerical techniques to evaluate the integral, which converges rapidly for increasing k. The result shows that in a cross section the average radial current is less than 2% of the average axial current if $x > 2a$. This inequality is readily satisfied within a small fraction of a space constant ordinarily.

ACKNOWLEDGMENTS

Preparation of this paper, including research on which new material is based, was supported in full by the Public Health Service through grants from the National Institutes of Health, HL23645 and HL31286.

REFERENCES

1. **Clark, J. W., Greco, E. C., and Harmon, T. L.,** Experience with a Fourier method for determining the extracellular potential fields of excitable cells with cylindrical geometry, *Crit. Rev. Bioeng.*, 31, 1, 1978.
2. **Clark, J. and Plonsey, R.,** A mathematical evaluation of the core conductor model, *Biophys. J.*, 6, 95, 1966.
3. **Clark, J. W. and Plonsey, R.,** Fiber interaction in a nerve trunk, *Biophys. J.*, 11, 281, 1971.
4. **Davis, L., Jr. and Lorente de Nó. R.,** Contribution to the mathematical theory of the electrotonus, in *A Study of Nerve Physiology*, Rockefeller Institute for Medical Research, New York, 1947, chap. 9.
5. **Geselowitz, D.,** Comment on the core conductor model, *Biophys. J.*, 6, 691, 1966.

6. **Hodgkin, A. L.,** *The Conduction of the Nervous Impulse*, Charles C Thomas, Springfield, Ill., 1964.
7. **Hodgkin, A. L. and Rushton, W. A.,** The electrical constants of a crustacean nerve fiber, *Proc. R. Soc. London Ser. B.*, 133, 444, 1946.
8. **Klee, M. and Plonsey, R.,** Integral equation solution for biopotentials of single cells, *Biophys. J.*, 12, 1676, 1972.
9. **Klee, M. and Plonsey, R.,** Extracellular stimulation of a cell having a non-uniform membrane, *IEEE Trans. Biomed. Eng.*, 21, 452, 1974.
10. **Klee, M. and Plonsey, R.,** Stimulation of spheroidal cells. The role of cell shape, *IEEE Trans. Biomed. Eng.*, 23, 347, 1976.
11. **Plonsey, R.,** *Bioelectric Phenomena*, McGraw-Hill, New York, 1969.
12. **Plonsey, R.,** The active fiber in a volume conductor, *IEEE Trans. Biomed. Eng.*, 21, 371, 1974.
13. **Plonsey, R.,** Generation of magnetic fields by the human body, theory, in *Biomagnetism*, Erne, S. N., Ed., Walter de Gruyter, Berlin, 1981.
14. **Plonsey, R. and Collin, R. E.,** *Principles and Applications of Electromagnetic Fields*, McGraw-Hill, New York, 1961.
15. **Plonsey, R. and Collin, R.,** Electrode guarding in electrical impedance measurements of physiological systems — a critique, *Med. Biol. Eng. Comput.*, 15, 519, 1977.
16. **Plonsey, R. and Heppner, D. B.,** Considerations of quasi-stationarity in electrophysiological systems, *Bull. Math. Biophysics.*, 22, 657, 1967.
17. **Van Den Honnert, C. and Mortimer, J. T.,** A technique for collision block of peripheral nerve: single stimulus analysis, *IEEE Trans. Biomed. Eng.*, 28, 373, 1981.

Chapter 3

CURRENT DENSITY MEASUREMENT IN LIVING TISSUE

W. B. Jarzembski

TABLE OF CONTENTS

I. ABSTRACT

Electrical stimulation for therapeutic purposes has been suggested since the time of Scribonius Largus (50 AD), but widespread application is just beginning. Part of the reason for such slow acceptance has been the lack of knowledge as to the mode of action of the electric currents. It is difficult to determine modes of action if the current pathways are unknown. For example, it has been hypothesized that all of the currents applied to the head pass through the integument, that the currents pass in a small tube directly between the two electrodes supplying the current, and that the current takes some intermediate pathway. It is clear that a better understanding of the current pathways is necessary if we are to develop an understanding of the mode of action of electric currents applied to the body for therapeutic purposes. It is the purpose of this chapter to discuss the means for experimental determination of current distributions in living tissue. The theories of direct, loaded probe method and the indirect, three and four electrode methods are elaborated.

II. HISTORICAL BACKGROUND

For almost 2000 years, attempts have been made to use electrical currents for therapeutic and/or analgesic purposes. For most of this period, progress in this field was impeded by lack of equipment to generate and apply the required currents as well as by a lack of understanding of the current pathways. For the past decade, design concepts and electronic apparatus have been available to provide any wave shape to any part of the body through surface or implanted electrodes. However, current pathways through the tissue remain in doubt.

III. THEORETICAL CONSIDERATIONS OF ELECTRICAL FIELDS IN NONHOMOGENEOUS MEDIA

Electrical field theory has been well developed for isotropic, homogeneous media with boundaries at infinity.[5] Field theory is also well developed for finite fields of specific geometries with a very limited number of media changes.[12,14] Using the assumption that there are no more than three or four different types of tissue, with respect to electrical properties, and that these different types are each located in homogeneous areas, analyses have been made of current pathways through the head for external electrodes.[3,13] These analyses have provided information with respect to the pathways of current within the brain and with respect to the ratio of current passing through the brain to that passing through the integument. Studies by this author, however, have shown that the inhomogeneous and anisotropic nature of the brain tissue may cause changes in current density of 1000% within 1 or 2 mm.[8] This is due to the fact that the electrical properties of the brain are anything but homogeneous and vary continuously throughout the entire brain with the exception of the ventricular system. The current density within various brain structures cannot be determined by means of classical field theory. Although finite element methods may yield insight, experimental measurements are required for definitive evaluation of current distributions.

There are two basic means available for the determination of current density. The electric field and local tissue resistivity may be determined for a specific location and the current density calculated from the relationship

$$J = E/\rho \qquad (1)$$

where J is current density, E is electric field, and ρ is resistivity. Due to the fact that the tissue properties change in relatively short distances, it is necessary to determine the electric

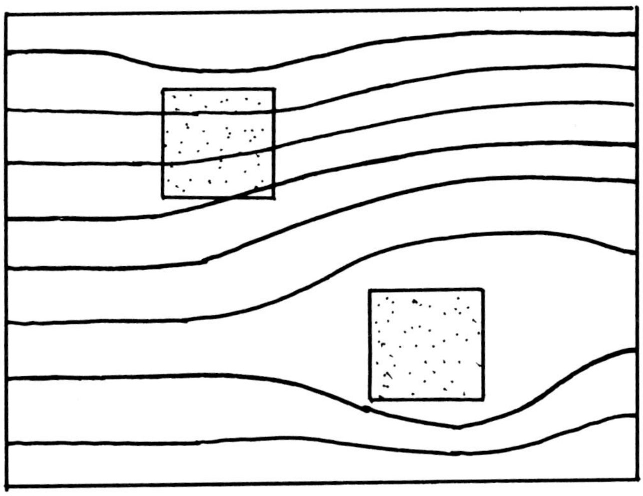

FIGURE 1. Effect of impedance anomalies on current pathway in volume conductor.

field strength and resistivity for the local area in which it is desired to determine the current density. The one other viable method for determining current density involves interrupting the conducting medium at a point and allowing the current to pass through a substitute circuit. Each of these methods will be described in detail.

A third method that has been suggested for the determination of current density is the hall effect method. This method relies on the fact that an electric current has associated with it a magnetic field. The hall effect device makes use of the fact that currents through a semiconductor will drift in a magnetic field, changing the electric field symmetry within the semiconductor. A pair of electrodes is mounted on the semiconductor device so as to be on an isopotential line when the magnetic field is essentially zero. Any increase in magnetic field strength will cause a drift of the current carriers, changing the shape of the electric field so that the two electrodes are no longer on an isopotential and a potential appears between the two electrodes. This is a proven concept and may indeed be used to determine the magnetic field strength at a specific location. Two problems that are associated with the use of such a device for the determination of current density within living tissue are the relatively large size of these devices and the fact that the devices are affected by all magnetic fields so that a large distal current will appear identical to a smaller local current.

Although an exact solution by mathematical methods is not possible for real tissue with the complexity of the brain, an understanding of the effects of nonhomogeneous inclusions will allow graphical approximations of current density fields. Figure 1 illustrates the distortion that will occur in an electric field due to embedded bodies with conductivities that are higher and lower than the conductivity of the surrounding medium. The lines represent current flow lines. It is clear from this plot that inclusions with low conductivity (high resistivity) cause a crowding of current into other available conducting materials, while inclusions of high conductivity will increase the local current density at the expense of adjacent current density.

The classical method of field plotting, the method of curvilinear squares, has been suggested for sketching fields to determine current densities and pathways. This method has been used by the author to sketch electrical fields to accuracy of better than 5% as compared to an analytical solution. Unfortunately, the method of curvilinear squares which is based on a graphical solution to the Laplace equation is applicable only to a two-dimensional field or a three-dimensional field that is unchanging in one dimension such as the field around a long, uniform cross section.

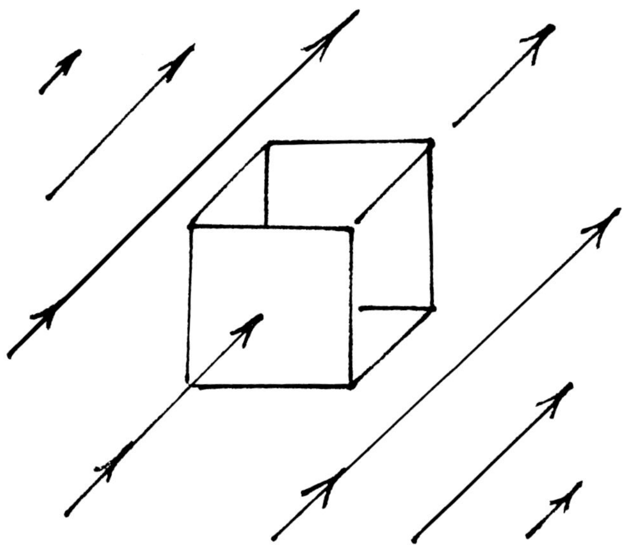

FIGURE 2. Hypothetical cube in electrical field.

Relaxation methods developed from the theory of finite differences may be applied to the solution of homogeneous electric field problems. This method is capable of yielding quite accurate graphic models in two dimensions with relatively little effort on the part of the investigator. By combining relaxation methods with topographical mathematical methods, it is theoretically possible to develop relaxation methods that are capable of three-dimensional solutions to electric field problems involving anisotrophic as well as nonhomogeneous media. Recently developed finite element models of the spinal cord yield good agreement with published current density values.[1,18,20] These methods require an extensive knowledge of the resistivity field within the media which is presently not complete.

IV. METHODS FOR MEASUREMENT OF CURRENT DENSITY

As previously stated, all methods proposed to date fall into two classes: (1) direct method where a device is inserted into the tissue so as to cause the local currents to pass through the device where it is measured and (2) indirect method where local resistivity values are obtained followed by a determination of the local electrical field. Current density is then calculated. Each of these methods is discussed in detail below.

A. Direct Measurement of Current Density

1. Cube Substitution Method

This method of current density and specific impedance was developed by the author[6] and subsequently used by Swiontek et al. for investigation of current distribution in spinal cord and cerebellum for various electrode configurations.[21] This method, which might be called the substitution method, makes use of an electrode of closely controlled geometry comprising an insulating body, essentially cuboidal in shape, having platinum electrodes on two opposing sides. The operation of this device may be explained using Figure 2 as a reference.

The theoretical basis for this current density measuring system is derived from classical electrical field theory. Imagine a small cube within a homogeneous electrolyte in a uniform electric field. The cube will be aligned with two faces parallel to isoelectric lines. If these two faces are made conducting, there will be no change in the electrical field and current will flow into one of the conducting faces, through the interior of the cube, and leave from

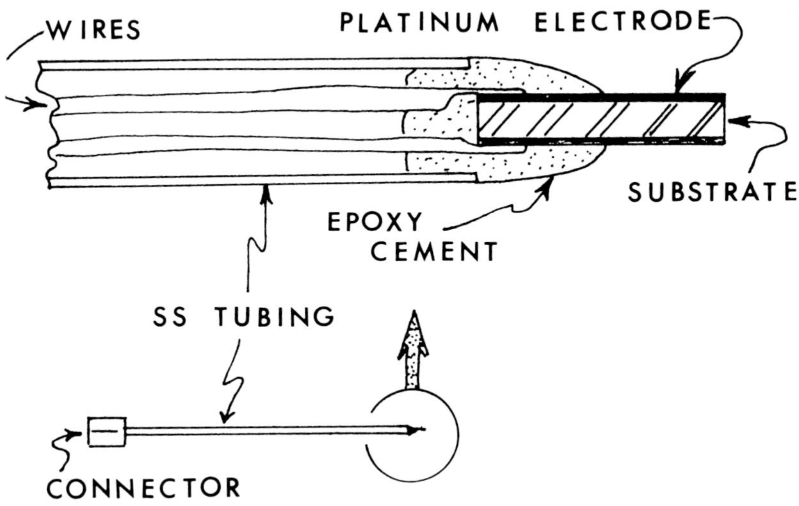

FIGURE 3. Sectional view of typical current density probe with parallel plates of platinized platinum.

the opposite conducting face. The cube may be viewed as a two-port device for which a Thevenin equivalent circuit may be derived. It should be clear that there are no internal sources of current or potential so the Thevenin equivalent potential will be zero. There are no active devices or energy storage elements within the cube so the interior may be viewed as a simple resistor. Thus, if the electrolyte within the cube is replaced with an appropriate resistance, there should be no change in the current flow or the external electric field.

The theoretical value of this resistance may be determined from the formula:

$$R = \rho L/A \tag{2}$$

where ρ is resistivity, A is cross-sectional area, R is resistance, and L is path length. If the cube has 1-cm sides, it is obvious that the resistance in ohms required for zero field distortion will be equal in value to the electrolyte resistivity in Ohm-centimeters (Ω-cm). A more practical size might be a cube with sides of 0.5 mm which would require a resistance of 20 times the electrolyte resistivity.

A practical embodiment of this device is shown in Figure 3, in which a slab of ceramic has been mounted on the end of a 10-cm length of type 302 SS hypodermic tubing by means of epoxy cement. Two faces of the ceramic have been coated with pure platinum and connecting wires have been carried through the barrel of the tubing. During assembly the platinum surface has been coated with an insulating dielectric except for the active electrode area. The active electrode area is then platinized by the Kohlrausch method.

It should be clear from the foregoing that there is a fixed ratio of electrolyte resistivity and internal cube resistance if the electric field is to remain undistorted. If the electrolyte resistivity is known, it is easy to calculate the resistance size for a known electrode system geometry. However, if the electric field is unknown and the electrolyte resistivity is unknown, it is necessary to use experimental methods to determine this relationship for each electrode system embodiment. This is a linear system and the properties of such systems allow a relatively simple calibration method to be used.

Each electrode system, hereinafter called "probe", may be individually calibrated with a test device (see Appendix) that provides a known uniform electrical field in a standard electrolyte. Electrolytes and field parameters may be changed to validate system linearity. The test device can provide an electric field strength measurable to within a fraction of a

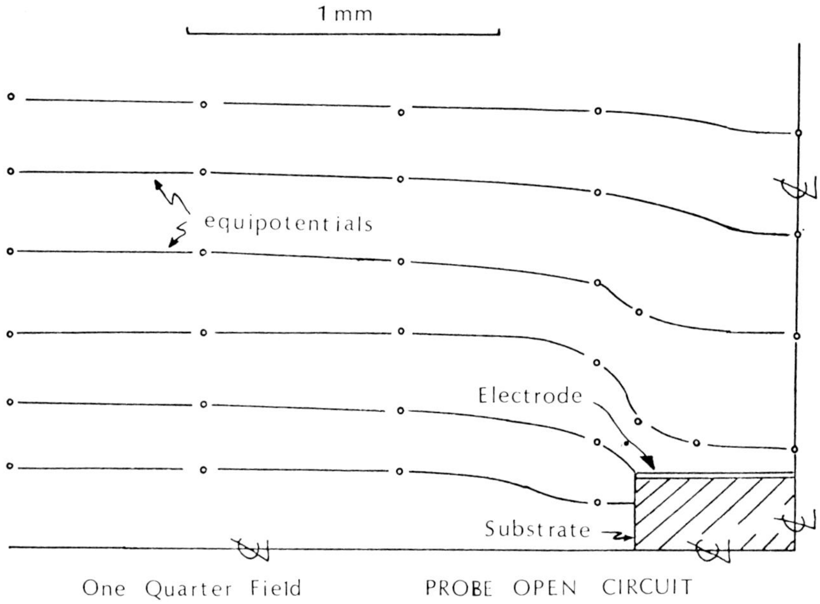

FIGURE 4. Experimentally determined electric field in vicinity of electrode with infinitely large (open circuit) load resistance.

percent. It may also make measurement of electrolyte resistivity to within the same accuracy. Physical dimensions of the probe may be measured quite accurately except for the actual active area of the electrodes. It is difficult to physically measure this area to an accuracy of 1%.

The actual calibration of an electrode starts with the measurement of the potential between the two electrodes while the probe is inserted in the electrolyte in a known electric field. This potential is first measured with no resistor in the probe circuit (infinite resistance). The resulting field is shown in Figure 4 which was obtained from measurements in the electrolyte surrounding a 1-cm cube. The external resistance is then lowered until the measured potential is approximately equal to the value predicted from the geometry of the probe and the known electric field strength. This value will also approximate the value calculated by use of Equation 2. It is well, at this point, to repeat this set of measurements with electrolytes of differing resistivity and with different electric field strengths.

The resistance that allows the probe to be inserted without field distortion will be referred to as the matching resistance. The ratio of the open circuit probe potential to the probe potential with matching resistance in the probe circuit will remain the same regardless of the electric field or electrolyte in which the probe is placed. This is true because of the linearity properties. It is, therefore, possible to make use of this ratio to determine the appropriate value of the matching resistor when the electric field and electrolyte resistance are unknown. When the external probe resistor is of the proper value to produce matched conditions, there is no field distortion and the current flowing through the matching resistor will be the same as the current that would flow through the volume occupied by the probe if the probe were not there.

One potential source of error would be the aforementioned electrode/electrolyte interface impedance as this impedance will be directly in series with the matching resistance. This impedance has been measured for many probes has caused less than 1% error when tested in electrolytes with 1000-Hz resistivities ranging from 50 to over 10,000 Ω-cm.

Another important source of potential error is the alignment of the probe so that the electrode faces are parallel to isopotential surfaces. A practical design will restrict rotation

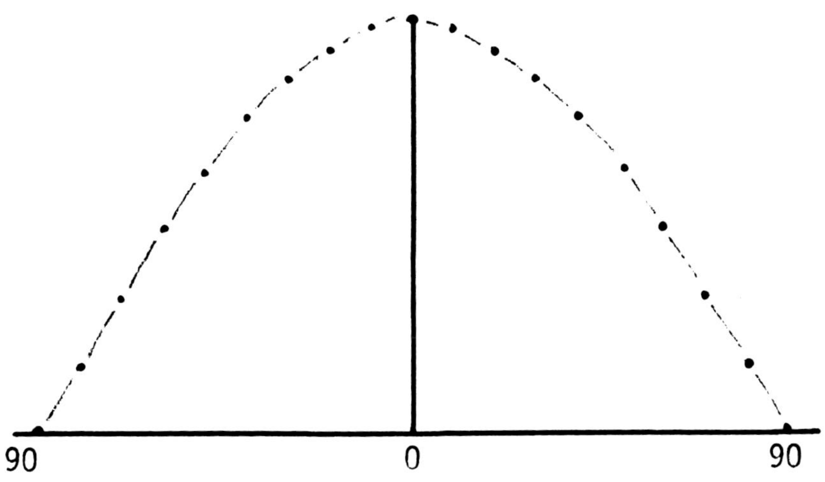

FIGURE 5. Probe output potential as a function of rotation about probe axis.

to one axis. It will, therefore, be necessary to make two measurements at each point with the probe inserted on orthogonal axes. The effect of rotation will be sinusoidal as shown in Figure 5.

In practice, the probe is positioned in the tissue with the tip at the location where current density measurement is desired, and the probe is rotated to the point of maximum open circuit potential. This potential is noted and loading resistance is placed in the external probe circuit to reduce the measured potential so that the ratio of open circuit potential to loaded potential is equal to the ratio for matched loading determined when the probe was calibrated as described above.

When this condition is achieved, it will be possible to measure the probe potential, the loading resistance, and the current through the loading resistance. These parameters will allow the calculation of the vector projections of electric field strength and current density perpendicular to the plane of the electrode surface. The electrical field vector will equal the value of the electrode potential divided by the electrode spacing; the current density vector will be equal to the current through the probe loading resistor divided by the effective electrode area; and the resistivity may be determined by reference to Equation 2 previously discussed.

The electronic circuitry for the above determinations may be based on the simplified circuit of Figure 6. Only two connections are required between the probe and measuring circuits. Good shielding and grounding practices will be most helpful in reducing interference. The potential measuring portion of the circuit requires only a low-noise, low-drift, high-gain isolated amplifier that may be tuned to the stimulation frequency. The current measurement is made by means of a low-noise, low-drift operational amplifier operated in the current mode. A suitable amplifier will introduce less than a milliohm of impedance in series with the loading circuit. The loading resistor may be a simple variable resistance for manual operation or a junction FET for automated operation. The stimulating frequency will allow the use of isolation transformer coupling.

Typical results of a manually operated system are shown in Figure 7. These results were obtained with a macaque monkey using a stimulation frequency of 2 kHz. Each reading was taken while advancing the probe into the brain and then again as the probe was withdrawn. These studies have shown current density variations as great as 1000% in less than 2 mm of probe travel. Studies to date have been preliminary to test the system, but the results indicate that the system is viable. Swiontek has also used this system for spinal cord study.[21]

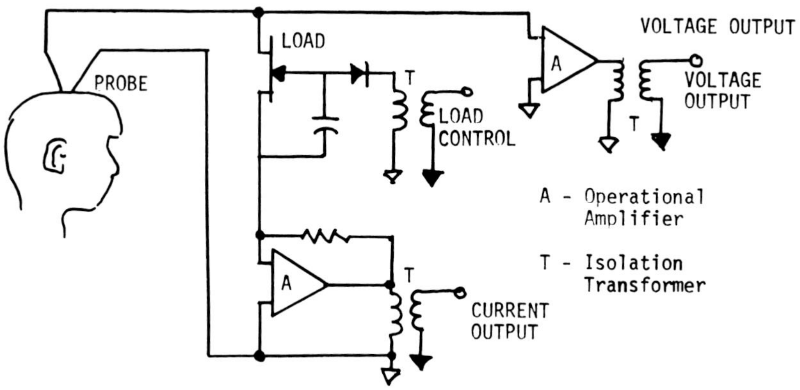

FIGURE 6. Current density measurement apparatus schematic diagram illustrating isolation method, control of load resistance, and measured values.

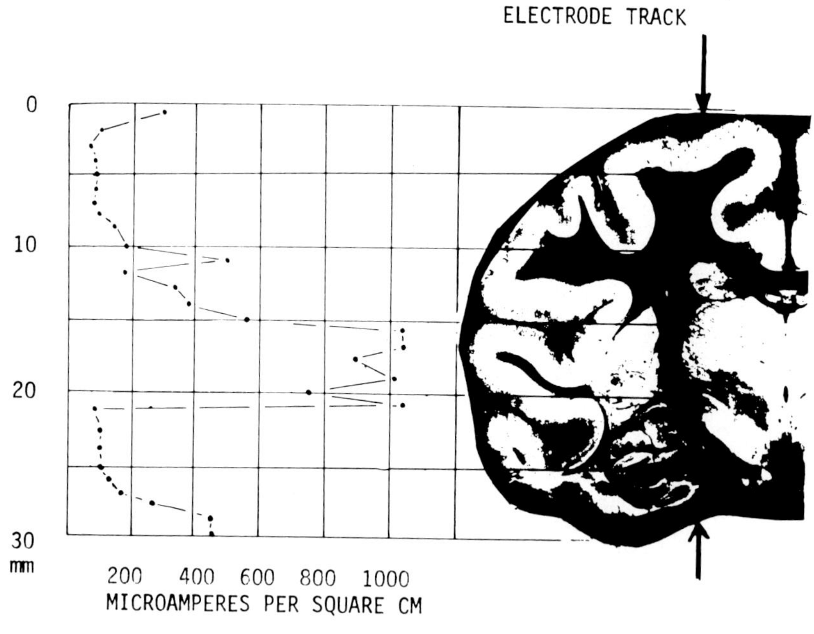

FIGURE 7. Current density vs. vertical position in brain of macaque monkey with 25-mA RMS applied current with glabella and inion at 1-kHz frequency.

2. *Three-Dimensional Electrode Array*

Deutsch[2] has proposed a chisel-pointed probe with four electrodes arranged so as to provide three-dimensional information relative to the electric field and current density. A tiny ceramic cylinder with a chisel point forms the structural support for the electrode array. Two of the four electrodes are spaced at opposite ends of the chisel point. The two additional electrodes are orthogonal to the edge of the chisel point and are set into the flat chisel sides so as to provide an axial offset. With these four electrodes, combinations may be chosen to provide information relative to the electric field along three orthogonal axes.

Deutsch proposes that current density information may be obtained for the three axes by loading each set of electrodes with an external resistance until the potential drops to one half of the open circuit potential. It is a well-known fact that loading to half potential provides maximum power transfer, but no evidence is presented to support the concept that such loading will result in zero field distortion or that such loading will provide the same current flow that would exist if the probe were not in position. It is difficult to see how current passing through a finite cross-sectional area could be the same as that passing through electrodes that occupy a small fraction of the area. This probe system could be useful, however, if experimental evidence were presented to show a relationship between current through the probe and the actual current density. A test apparatus such as that described in the Appendix could provide such information easily.

B. Indirect Methods of Current Density Measurements

Measurements of electric field strength and resistivity can be used to calculate current density by the relationship.

$$J = E/\rho$$

where J is current density, E is electric field strength, and ρ is tissue resistivity. For a homogeneous, isotropic medium in a uniform field, ρ is a scalar. For anisotropic media, ρ must be represented by a tensor. Electric field strength and tissue resistivity can be determined in several ways. The most commonly used are the three and four electrode methods.

1. The Three- and Four-Electrode Methods

The relationship between resistance and resistivity for a volume conductor, of uniform cross-sectional area, between two conducting electrodes is

$$R = \rho l/A \tag{3}$$

where R is resistance, ρ is tissue resistivity, l is length, and A is cross-sectional area. To determine the resistance of various thickness shells of spherical shape, we must account for the constantly changing cross-sectional area. This is done by considering a very thin shell and integrating

$$dR = \rho dr/4\pi r^2 \tag{4}$$

$$R = \frac{\rho}{4\pi}\int_{r_o}^{r_1} dr/r^2 = \frac{\rho}{4\pi}(1/r_o - 1/r_1)$$

To determine resistivity, a current, I, can be injected at point electrode in the medium and sinked distally. The potential difference between two points at distances r_o and r_1 from the current source is measured and divided by I to yield the shell resistance (R) between r_o and r_1. Resistivity of the shell media, ρ, is calculated from Equation 5

$$R = \frac{\rho}{4\pi}\left(\frac{1}{r_o} - \frac{1}{r_1}\right) \tag{5}$$

In practice it is often convenient to equally space the electrode for these measurements "a" units apart. Thus, resistance becomes

$$R = \frac{\rho}{4\pi}\left(\frac{1}{a} - \frac{1}{2a}\right) = \frac{\rho}{8\pi a}$$

for a uniform spherical shell.

If the current is injected at the surface of a flat volume conductor, the above equation becomes

$$R = \frac{\rho}{4\pi a}$$

The four-electrode method is a common modification of the three-electrode method in which the distal electrode is replaced by a fourth, equally spaced electrode. Current is applied between the outer and the voltage is measured between the inner two electrodes. Superposition indicates that the voltage measured with the four-electrode system will be twice that of a three-electrode system:

$$R = \frac{\rho}{4\pi a}$$ for electrodes deep within a medium and

$$R = \frac{\rho}{4\pi a}$$ for electrodes at the surface of a slat semi-infinite conductor.

The theoretical bases for all the above methods for determining resistivity and current density assume that the media is linear, homogeneous, and isotropic.

If the assumption of isotropicity is set aside, measuring ρ and J generally becomes exceedingly difficult in biological systems. In this case ρ becomes a tensor quantity relating the vectors E and J.

Some special anisotropic cases have been handled analytically as well as experimentally. In an anisotropic medium described by three principal resistivities along the axis of a three-dimensional coordinate system, closed form expressions can be derived to relate voltage distribution in the medium to current applied via electrodes at the boundaries.[16]

For example, if principal resistivities ρ_x, ρ_y, and ρ_z are in the x, y, and z directions of a conventional rectangular coordinate system, then the three-electrode or the four-electrode methods described above can be used to find ρ_x, ρ_y, and ρ_z. With the electrodes oriented in the x direction, current (I) is injected between the outer electrodes (one and four). Voltage ($V_{2,3x}$) between the inner electrodes (two and three) is proportional to the product of resistivities orthogonal to x.

$$\rho_y\rho_z = \frac{(V_{2,3x})^2 4\pi^2 a^2}{I^2}$$

If the voltage measurement is repeated in the y and z directions, the three resulting equations can be solved for ρ_x,ρ_y, and ρ_z. Again, if we consider a semi-infinite medium with two measurements made at the surface and the other made in depth, the resulting resistivities should be scaled by one half.

If the resistivity in depth (ρ_z) is the same as one of the other (ρ_y), then a single measurement in the x direction will yield the value for ρ_y and ρ_z. A subsequent measurement in the y direction will give ρ_x without an indepth measurement. This makes possible resistivity measurements in some anisotropic tissues with surface electrodes only. The region represented by the three- or four-electrode method may be determined using Equation 5. If r_o is the electrode radius and r is the radius of the spherical shell of tissue around the electrode, it can be shown that the total impedance between a small spherical electrode and a large distal electrode in a uniform, isotropic medium will be only 10% more than the impedance contributed by the medium within 5 diameters of the small electrode. Lang[10] has experimentally determined that the electrical field near a small electrode mounted at the distal end of a hypodermic needle is relatively uniformly distributed and that indeed such a system is uniform in estimating local current density. He has further made plots of potentials surrounding such small electrodes in primates. Equipotential surfaces in these plots are remarkably spheroid in shape, thus validating the use of this method as an indicator of the electric field and current density.

For electrode radius r_o, the following table gives values of R for spherical shells ranging from infinity to zero with an electrode diameter of unity:

Resistance, R	**Shell radius**
1.000	∞
0.999	1000 r_o
0.990	100 r_o
0.900	10 r_o

Thus, it can be clearly seen that 90% of the total resistance lies between the small spherical electrode and a shell that has a radius 10 times the radius of the electrode. For an electrode with a tip radius of 100 μm, it would appear that 90% of the resistance in the circuit lies within a sphere with a radius of 1 mm (1000 μm). This is based on the assumption that the distal electrodes are placed at a distance that is very large compared with the electrode radius. The experimental work of Lang[9] tends to justify this assumption.

IV. CONCLUSIONS AND RECOMMENDATIONS

This chapter demonstrates that the in vivo measurement of current density, with presently available techniques, is not a simple task. To develop reasonably accurate maps of current pathways suitable for analysis of the mode of action of electrical stimulation, it is necessary to make a large number of measurements. Even with the computerized direct measurement system proposed by Jarzembski,[9] this would be a tremendous task. The indirect methods, although using simpler measurement apparatus, will require even more effort.

It is, therefore, clear that development of a better method is highly desirable, although it is not presently clear from known theory what directions such development would take. In the meantime, it appears prudent to hypothesize that the observed results of electrical stimulation are the result of currents passing through a particular portion of the brain. It then becomes feasible to use one of the foregoing methods to investigate the current density in this particular structure.

Although not discussed in great details in this paper, it is considerably easier to map impedance in the tissue. Numerical relaxation and finite element methods should lead to a procedure for computerized calculation of currents in specific areas. Such models could be tested by making relatively few actual measurements. With the renewed interest in electrical stimulation for therapeutic purposes, it is important that research be carried on to determine accurately the structures being affected by these currents.

Although a great deal of work has been done to determine current densities in materials with uniform properties in recent years, few studies have been done to determine current pathways and current density in living tissue. If these new therapeutic modes are to see widespread acceptance, it will be necessary to continue the study of methods for current pathway and current density determination.

ACKNOWLEDGMENT

Support for the above was partially supplied by the National Institutes of Health under Grant 5 R01 GM 24104-03.

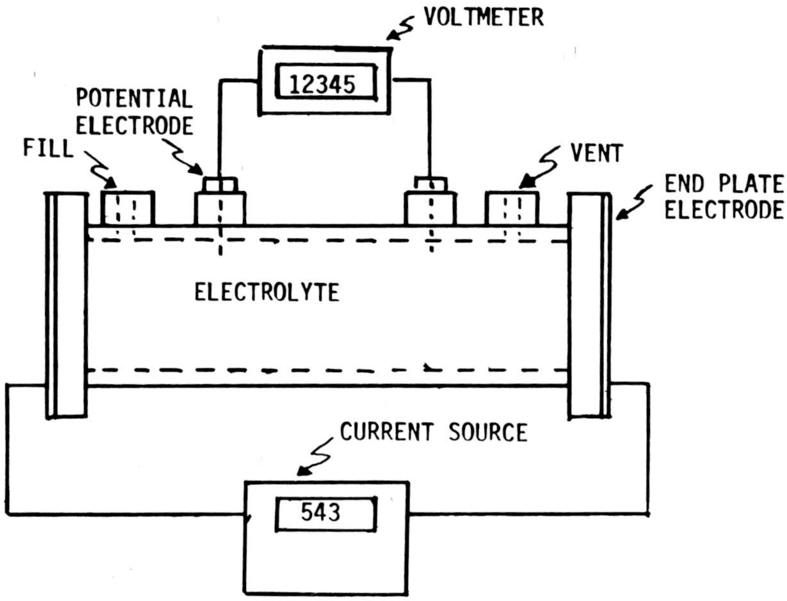

FIGURE 8. Resistivity test cell.

APPENDIX

Test Cell for Resistivity Measurement

The test cell shown in Figure 8 has been designed for the direct measurement of liquid electrolyte resistivity (specific impedance). The electric field within the cell is quite uniform which allows the cell to be used for calibration of certain types of electrode measuring systems that will be used for in vivo measurement of resistivity and current density. Two end plate electrodes are provided for passing current through the electrolyte. The design of the cell is such that the current distribution across the liquid volume is uniform. A pair of potential electrodes is located along the length of the liquid volume that may be used to determine the electric field strength. If the total current is known and the electric field is known, the resistivity may be determined.

The resistance of a volume with uniform cross section is given in Equation 6 below. The geometry of the test cell may be accurately determined by physical measurement so that length of the measured volume (between the potential electrodes) is known and the cross-sectional area (cross-sectional area of inner space) is known. The resistance may be accurately determined as the ratio of the measured potential to the measured current. Substitution of these three values in Equation 6 will result in the determination of resistivity. Resistivity varies greatly with temperature so provision is made for temperature measurement. This cell also provides potential and current density fields of uniformity suitable for calibration of current density measuring systems.

$$R = \rho L/A \tag{6}$$

where R is resistance, ρ is resistivity, A is cross-sectional area, and L is length.

REFERENCES

1. **Coburn, B.,** Electrical stimulation of the spinal cord: two-dimensional finite element analysis with particular reference to epidural electrodes, *Med. Biol. Eng. Comput.*, 18, 573, 1980.
2. **Deutsch, S.,** A probe to monitor electroanesthesia current density, *IEEE Trans. Biomed. Eng.*, 15, 130, 1968.
3. **Driscoll, D. A.,** An Investigation of a Theoretical Model of the Human Head with Application to Current Flow Calculations and EEG Interpretation, Ph.D. dissertation, University of Vermont, Burlington, 1970.
4. **Hayes, K. J.,** The current path in electric convulsive shock, *Arch. Neurol. Psychiatry*, 63, 102, 1950.
5. **Hayt, W. H., Jr.,** *Engineering Electromagnetics*, 2nd ed., McGraw-Hill, New York, 1967.
6. **Jarzembski, W. B., Larson, W. V., and Sances, A., Jr.,** Evaluation of specific cerebral impedance and current density, *Ann. N.Y. Acad. Sci.*, 170, 476, 1970.
7. **Jarzembski, W. B.,** Electrical Current Distribution in the Brain During Application of Diffuse Electrical Currents, Ph.D. dissertation, Marquette University, Milwaukee, 1971.
8. **Jarzembski, W. B.,** Cerebral current density measurements during application of electrical currents thru surface electrodes, in *Physiological Applications of Impedance Plethysmography*, Allison, R. D., Ed., ISA Press, Pittsburgh, 1972, chap. 3.
9. **Jarzembski, W. B.,** Current vector measurement apparatus under direct computer control, *Front. Eng. Health Care*, 1, 10.6.1, 1980.
10. **Lang, J., Sances, A., Jr., and Larson, S. J.,** Determination of specific cerebral impedance and cerebral current density during the application of diffuse electrical currents, *Med. Biol. Eng.*, 7, 517, 1969.
11. **Lorimer, F. M., Segal, M. M., and Stein, S. N.,** Path of current distribution in brain electroconvulsive therapy, *Electroencephalogr. Clin. Neurophysiol.*, 1, 343, 1949.
12. **Moon, P. and Spencer, D. E.,** *Field Theory for Engineers*, D Van Nostrand, New York, 1961.
13. **Plonsey, R.,** *Bioelectric Phenomena*, McGraw-Hill, New York, 1969.
14. **Plonsey, R.,** Electrical field modeling, in *Neural Stimulation*, Vol. 1, Myklebust, J. B., Sances, A., Jr., Larson, S. J., and Cusick, J. F., Eds., CRC Press, Boca Raton, Fla., 1985, chap. 2.
15. **Ranck, J. B., Jr. and Bement, S. L.,** The specific impedance of the dorsal columns of cat: an anisotropic medium, *Exp. Neurol.*, 11, 451, 1965.
16. **Rush, S.,** Methods of measuring the resistivities of anisotropic conducting media in situ, *J. Res. Natl. Bur. Stand.*, 66c(3), 217, 1962.
17. **Rush, S. and Driscoll, D. A.,** Current distribution in the brain from surface electrodes, *Anesth. Analg.*, 47, 717, 1968.
18. **Rusinko, J. B., Walker, C. F., and Sepulveda, N. G.,** Finite element modeling of potentials within the human thoracic spinal cord due to applied electrical stimulation, in *IEEE 1981 Frontiers of Engineering in Health Care*, IEEE, Piscataway, N.J., 1981, 76.
19. **Sances, A., Jr., Myklebust, J. B., Szablya, J. F., Swiontek, T. J., Larson, S. J., Chilbert, M., Prieto, T., Cusick, J. F., Maiman, D., and Pintar, K.,** Current pathways in high voltage injuries, *IEEE Trans. Biomed. Eng.*, 30(2), 118, 1983.
20. **Sepulveda, N. G. and Walker, C. F.,** Finite element modeling of potentials within the human brain due to applied electrical stimulation, *2nd IEEE-EMBS Conference*, IEEE, Piscataway, N. J., 1980.
21. **Swiontek, T. J., Sances, A., Jr., Larson, S. J., et al.,** Spinal cord implant studies, *IEEE Trans. Biomed. Eng.*, 23, 307, 1976.
22. **Weaver, L., Williams, R., and Rush, S.,** Current density in bilateral and unilateral ECT, *Biol. Psychiatry*, 11, 303, 1976.

Chapter 4

EXPERIMENTAL CURRENT DENSITY MEASUREMENT

Thomas Swiontek and Anthony Sances, Jr.

TABLE OF CONTENTS

I. INTRODUCTION

Knowledge of the current distribution secondary to neural stimulation is important in the interpretation of physiologic effects.[6,7,13,22] Furthermore, the experimental measurement of current density and tissue impedance makes possible mathematical models[1,12,19] and promotes the development of optimal stimulation systems.[18]

Spinal cord current distribution has been evaluated for three-in-line and five-in-line alternate polarity systems, a four-electrode parallel system, a system designed to pass current from anterior to posterior in the spinal cord.[4,7,18,20,21] Current density with cerebellar stimulation has been measured using alternate polarity systems,[3] systems designed to pass current from superior to inferior surfaces on the cerebellum, and for monopolar arrays.[14,15] Additionally, studies have been conducted to determine the current pathways secondary to externally applied currents.[17,18]

A parallel plate, 0.3 mm on a side, loaded probe was used to make current density and resistivity measurements in the cerebellum and spinal cord of monkeys and human cadavers[3,4,7,14,15,18,20,21] (Figure 1). The fabrication and calibration of the probe is similar to that of the larger probe used to measure cerebral current density secondary to transcranial current application.[2] Briefly, the larger probe consists of a ceramic cube with platinum-platinum black electrodes on two opposite faces. The cube is mounted on a small-diameter insulated steel shaft which permits insertion of the cube (tip) into brain tissue. Since the resistance of the ceramic is large, negligible current flows through the probe when it is placed in an electric field. This distorts the field near the probe tip. Placing the proper resistance in parallel with the platinum electrodes restores the original field pattern.

Since the shaft of the 0.3-mm probe could not be made much smaller than the tip size, the electric field near the tip could not be restored to an undistorted shape simply by loading. However, theoretically accurate measurements of current density could still be obtained under this condition by extending the theoretical basis underlying the loaded probe technique.[20]

The scaler potential, Φ, throughout a linear, isotropic homogeneous medium is uniquely determined (though it may be unknown) after specifying all boundary conditions. The scaler potential is any function which satisfies the boundary conditions and Laplace's equation[1] provided no sources of charge exist in the region.[9]

$$\nabla^2\Phi = 0 \tag{1}$$

Φ is scalar potential and ∇ is defined as

$$\left(\bar{\alpha}_x \frac{\partial}{\partial x} + \bar{\alpha}_y \frac{\partial}{\partial y} + \bar{\alpha}_z \frac{\partial}{\partial z}\right)$$

where $\bar{\alpha}_i$ is the unit vector in the "i" direction.

Equation 2 relates current density ($\bar{J}$), at any point in the medium, to Φ

$$\bar{J} = \frac{-\nabla\Phi}{\rho} \tag{2}$$

where $\bar{J}$ is given in amperes per square centimeter (A/cm²), $\nabla\Phi$ in volts per centimeter (V/cm), and ρ is the resistivity of the medium in (Ω-cm).

When the current density probe is in a homogeneous conductive medium with a uniform electrical field, E, of large extent relative to probe size, the boundary conditions which determine the solution, Φ, are

1. Probe geometry
2. Probe orientation with respect to the field

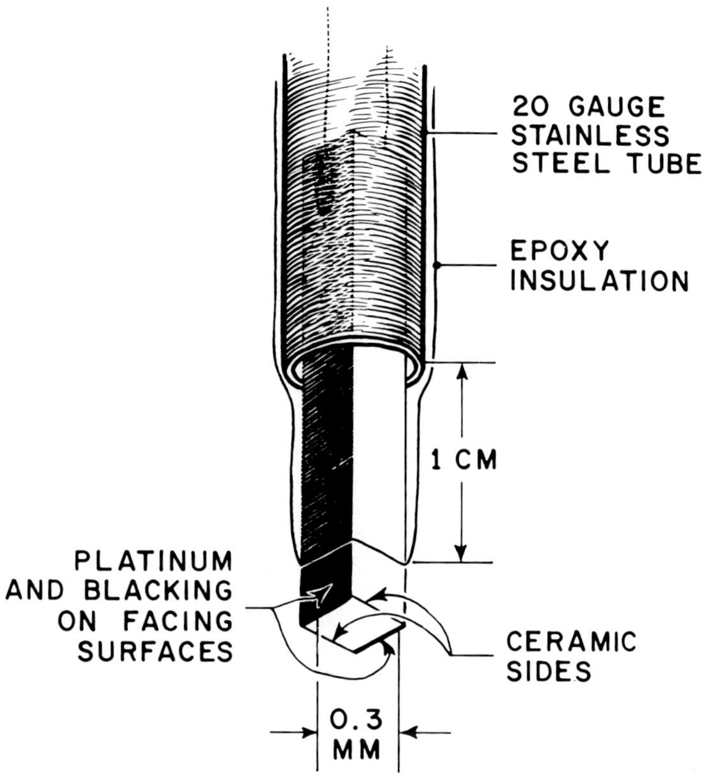

FIGURE 1. Details of the probe used to measure current density. Requirement for a narrow but rigid probe is satisfied by a ceramic substrate coated with platinum.[20]

3. Magnitude of the applied field, E
4. Voltage between conductive surfaces of the probe, V

This assumes that only insulators or conductors are in electrical contact with the medium.

An extended field problem can be considered by restatement of boundary condition (4) above as follows:

$$V = IR_L \tag{3}$$

where R_L is the external load resistor on the probe and I is current through R_L. "I" can also be expressed as

$$I = \int_S \bar{J} \cdot d\bar{s} \tag{4}$$

where S is the surface of the probe face and $d\bar{s}$ is a differential surface area with vector direction orthogonal to the surface.

Equation 5 is derived by combination of Equations 2 to 4.

$$V = \frac{R_L}{\rho} \int_S - \nabla\Phi \cdot d\bar{s} \tag{5}$$

If Φ could be found, given probe voltage, V, and field intensity, E, a series of calibration curves could be generated to relate ρ to R_L and V from Equation 5. Unfortunately, it is

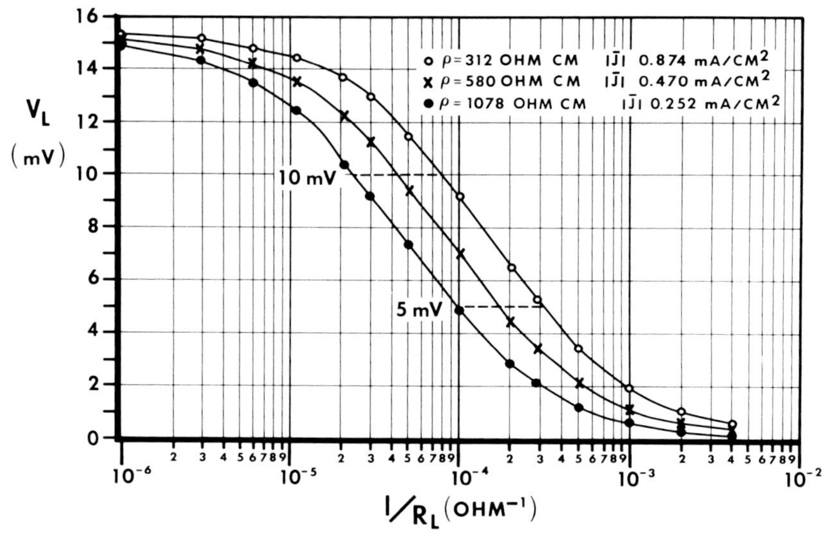

FIGURE 2. Voltage across a loaded probe is a monotonic function of the reciprocal of the load resistance. Curves are identical and spaced along the abscissa by an amount corresponding to the ratio of the resistivities of the media. Applied field intensity E was 0.27 V/cm in each.[20]

difficult to obtain analytical forms for the scalar potential, Φ, in the region of the current probe used in these experiments.

However, the relation between V, ρ, R, and E can be found empirically if boundary conditions 1 and 2 are held constant. Figure 2 shows voltage measured between the conductive surfaces as a function of R_L for one of the probes used in these experiments. For comparison, three different media were used, but with the same field intensity, E, in each. Since the solution, Φ, is directly proportional to E, values shown in Figure 2 can be appropriately scaled for any field intensity in a linear medium. As load resistance approaches infinity (unloaded condition), voltages measured across the probe in all media asymptotically approach the same value (approximately 15.5 mV in this case) which is a function of field intensity only. Moreover, it is found that if the voltage is reduced by some fixed ratio (K) of the unloaded voltage, V_u, the current which flows into the probe will be related to current density in the medium at large by a constant. For example, Figure 3 shows that if the probe voltage is reduced to 65% (K = 0.65) of its unloaded value, 0.9 μA of current will flow in the probe for each milliampere per square centimeter (mA/cm² of current in the medium. If the probe voltage is reduced to 32% of its unload value 1.8 μA of current will flow in the probe for each mA/cm² of current density in the medium. Thus, the choice of K (probe factor) to be used with a given probe is arbitrary. Probe factors used in these experiments were chosen so that average current density at the probe tips under load would be about the same as that of the medium.

Each probe is calibrated individually by immersion into a medium having a known, spatially uniform current density. Open circuit voltage, V_u, is recorded and the probe loaded until current flowing through the external load equals current density in the medium times cross-sectional area of the probe face. Voltage, V, across the probe at this load is recorded and the probe factor K determined by the ratio of these voltages: $K = V_L/V_u$. Medium resistivity can also be found with the loaded probe technique. Equation 5 indicates that probe voltage, v, depends upon the ratio of ρ to R_L rather than each independently. This is verified by Figure 4 in which data of Figure 3 are replotted with ρ/R_L as the abscissa. For each K, then, there is a unique value for ρ/R_L. Therefore, if R_L and K are known, ρ can be deduced.

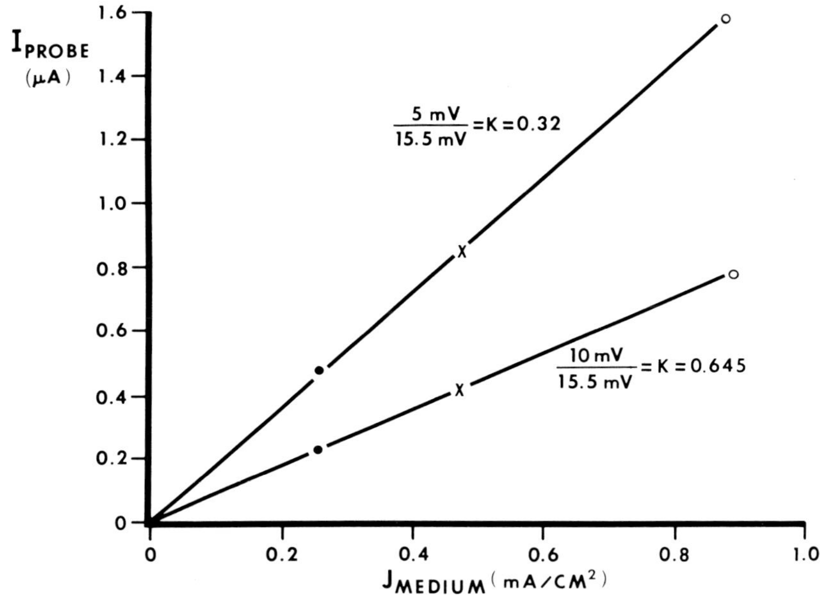

FIGURE 3. Loaded probe current is directly proportional to the magnitude of the current density in the surrounding medium when the probe voltage is a fixed ratio of the unloaded voltage. Data were derived for load voltages of 5 and 10 mV from Figure 2 as examples; linear relation holds over wide range of load voltage.[20]

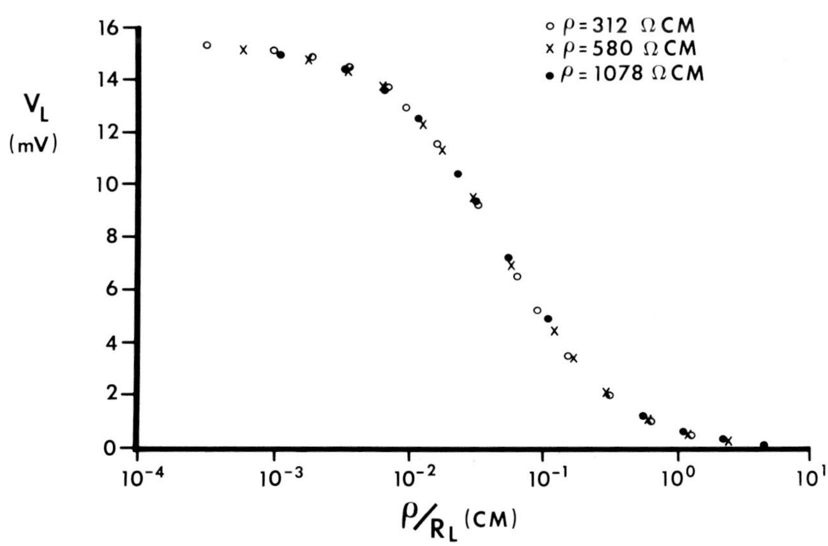

FIGURE 4. Voltage between probe surfaces under varying load conditions. Probe voltage is a function solely of medium resistivity/load resistance when magnitude and orientation of the applied electric field are held constant.

II. SPINAL CORD STUDIES

Electrodes from commercially available subdural spinal implant systems (Figure 5) were positioned on the dorsal columns of the spinal cord. In the case of the anterior-posterior

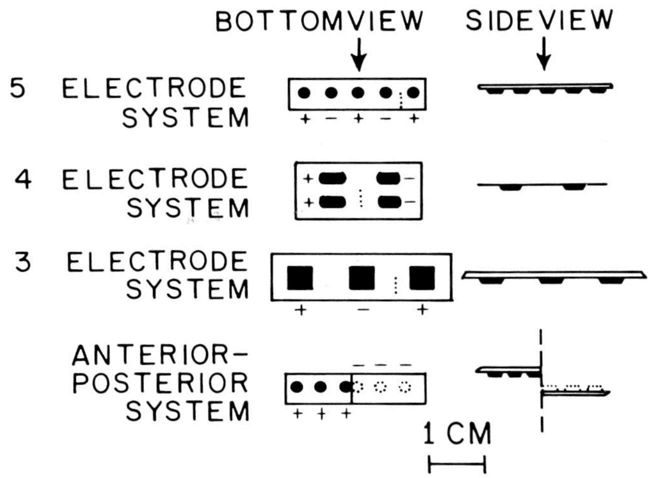

FIGURE 5. Electrode configurations for spinal implant systems. Electrodes shown with like polarities are electrically connected together in electrode arrays constructed for the experiments just as they are in the implanted units. Small dots (:) indicate points where probe was inserted.[20]

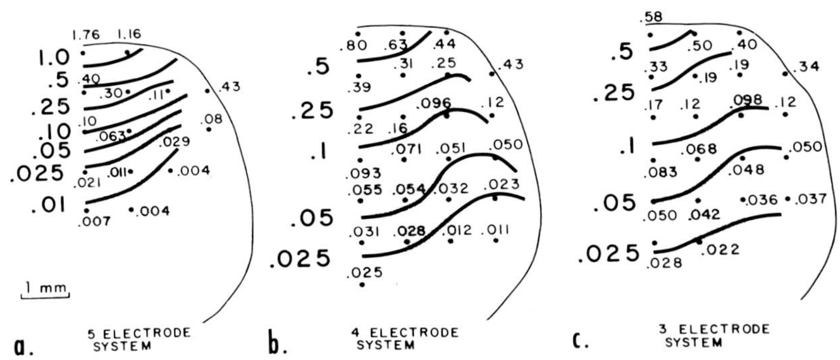

FIGURE 6. Current density in cadaver cord (mA/cm^2). Current density readings have been superimposed over a tracing of an enlarged photograph of a cross section of a cadaver cord used in the experiments. Current flow is essentially orthogonal to the cross section at points shown. Boldface lines were handdrawn between the points showing actual measurements.[20]

system, three negative electrode discs were placed over the ventral columns and three positive discs were placed over the dorsal columns.

The probe measures only the current normal to the probe surface. It does not detect current components parallel to the probe faces or along the axis of the probe shaft. Electric field plots on teledeltos paper with subsequent confirmation in a few spinal cords indicated that current flow was largely parallel to the axis of the spinal cords at points of equal distance from positive and negative electrode discs. Therefore, the probe was inserted midway between electrodes (Figure 5) with the faces transverse to the spinal cord axis. Measurements were made at 1-mm depth and 1-mm lateral spacings from the cord midline.

The current density distributions resulting from the application of current to fresh human cadaver cords via the dorsal column electrode systems are shown in Figure 6. Numerical integration of the current from Figure 3 indicates that the five-electrode system (five-in-line electrode system) confines 75% of the total applied current to the cord. The four-electrode and three-electrode systems distribute 22 and 40%, respectively, of the total applied current to the cord. Similar findings were observed using a saline model.

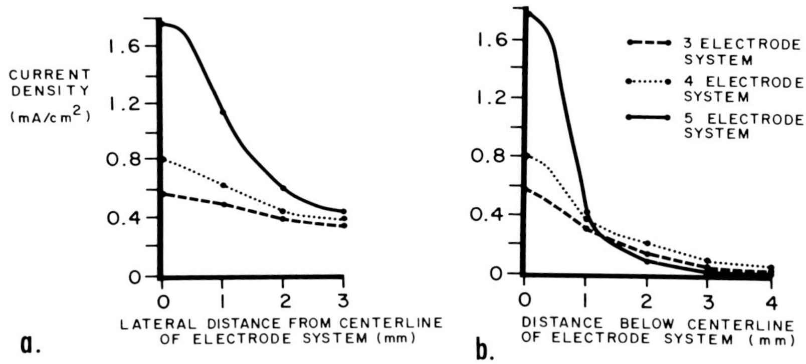

FIGURE 7. Current density in fresh cadaver cord. (a) As a function of distance lateral to the midsagittal plane along the surface of the cord; (b) as a function of depth from the dorsal surface in the midsagittal plane.[20]

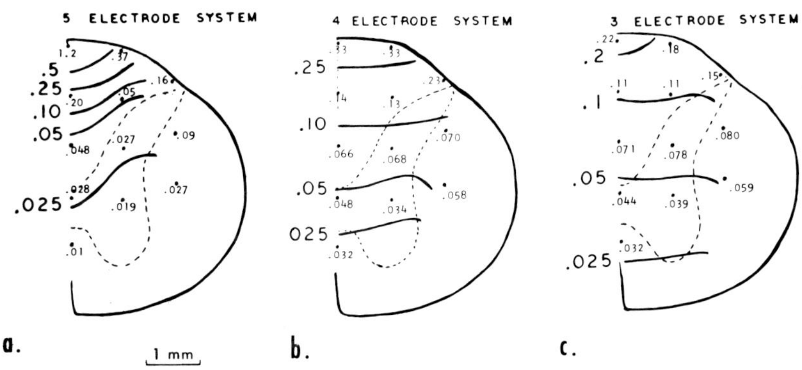

FIGURE 8. Current density readings (mA/cm²) from in vivo monkey spinal cord have been superimposed over a tracing of an enlarged photograph of one of the monkey cord cross sections. Current flow is orthogonal to cross section at points shown.[20]

The five-electrode system has the highest current density in the region near the electrodes. However, the current decreases rapidly away from the electrode so that at 3 to 4 mm lateral to midline (the width of a human spinal cord), it is approximately equal to that of the other two systems (Figure 7). Current density for the 5-electrode system also decreases more rapidly with depth (Figure 7). The depth and lateral current distributions of the three- and four-electrode systems are similar (Figure 7).

Comparison of the current distributions from the dorsal column electrode systems in live monkey cords (Figures 8 and 9) with the cadaver cord experiments shows the same relative differences between electrode systems. The magnitude of the current densities at each of the points measured in monkey cord are lower than at the corresponding points in cadaver cord. This result is expected since the smaller monkey cord allows more current shunting into the saline bath than the human specimens.

A markedly different current distribution results from the anterior-posterior system which is designed to pass current across the cord. Near the center of the electrode system (Figure 10) current passes at an angle of about 60° with respect to the long axis of the cord, or nearly directly from the anterior to the posterior electrode set. At a distance 3 to 7 mm from the center of the electrode system, current passes almost directly across the cord; further along the centerline of the cord the current again travels obliquely across it but away from

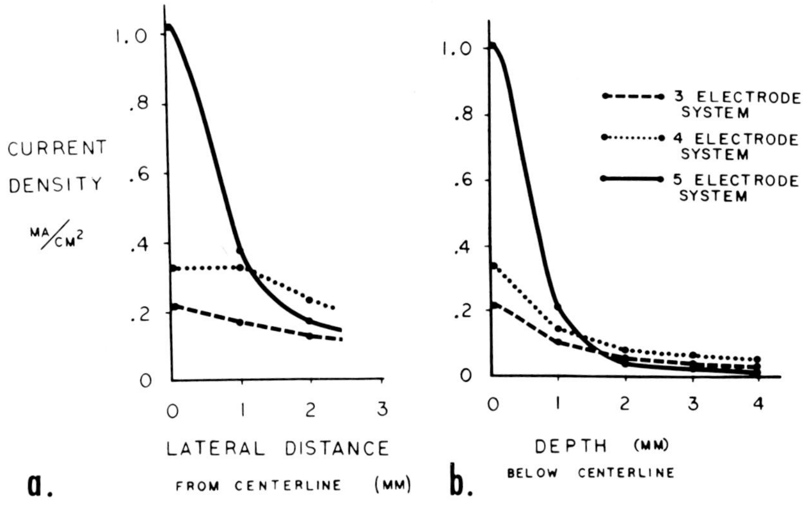

FIGURE 9. Current from in vivo monkey spinal cord. (a) As a function of distance lateral to the midsagittal plane along the surface of the cord; (b) as a function of depth below the dorsal surface in the midsagittal plane.[20]

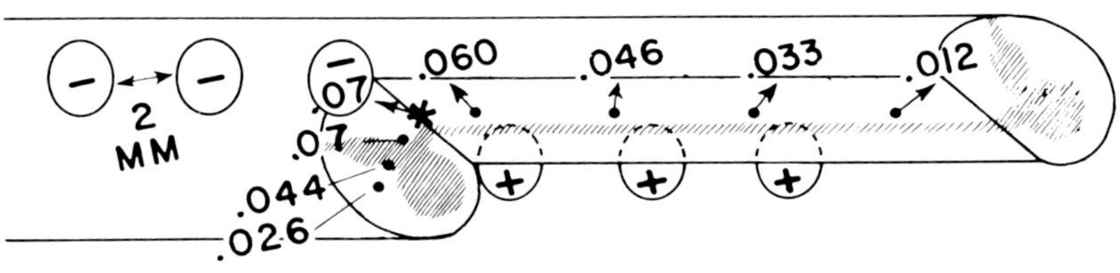

FIGURE 10. Current density distribution (mA/cm^2) in cadaver spinal cord from anterior-posterior electrodes. Current density vectors shown lie in midsagittal plane. Current flow at points lateral to the center (*) of the anterior-posterior system is parallel to vector at *.[20]

the electrodes. With the anterior-posterior (A-P) system the highest current densities are found in the midsagittal plane near the electrodes. The currents decrease rapidly with distance lateral to the center of the cord, similar to the five-in-line system.

Spinal cord resistivity was determined in six spinal cords using plexiglass chambers (Figures 11A, and B).[17] Transverse impedance was measured by passing 10-μA to 5-mA, 60-Hz currents between the stainless steel side plates (Figure 11A). Voltage was measured between electrodes adjacent to the plates. Longitudinal impedance of the spinal cord was found with the system of Figure 11B and used 3-in-line 0.25-cm spaced Nichrome electrodes insulated except at the tip for potential measurements. The center electrode was used as a reference. The current was passed from the cut ends of the cord lying in the trough. The mean transverse resistivity was 1970 Ω-cm with a SD of 1055. The mean longitudinal resistivity was 214 Ω-cm and the SD was 46.7.

III. CEREBELLAR STUDIES

Since the majority of clinical studies have been done with alternate polarity arrays.[5,6] measurements were made using this system in the fresh human cadaver cerebellum and in living monkeys.[3] Figure 12 illustrates the location of the electrodes on the human cerebellum and the measurement site. Figure 12B represents a coronal section showing the electrode

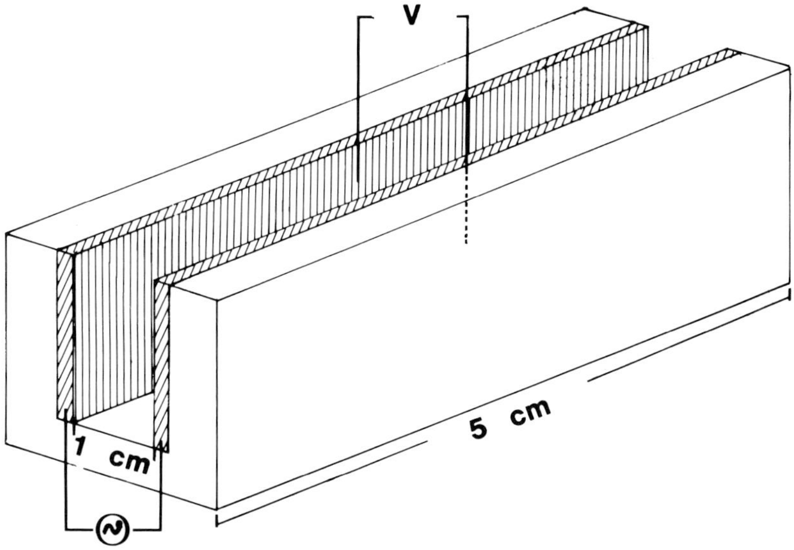

FIGURE 11A. Chamber in which transverse spinal cord resistivity was measured. Shaded region shows stainless steel current application electrodes. Voltage (V) measured with Nichrome wires in the tissue.[17]

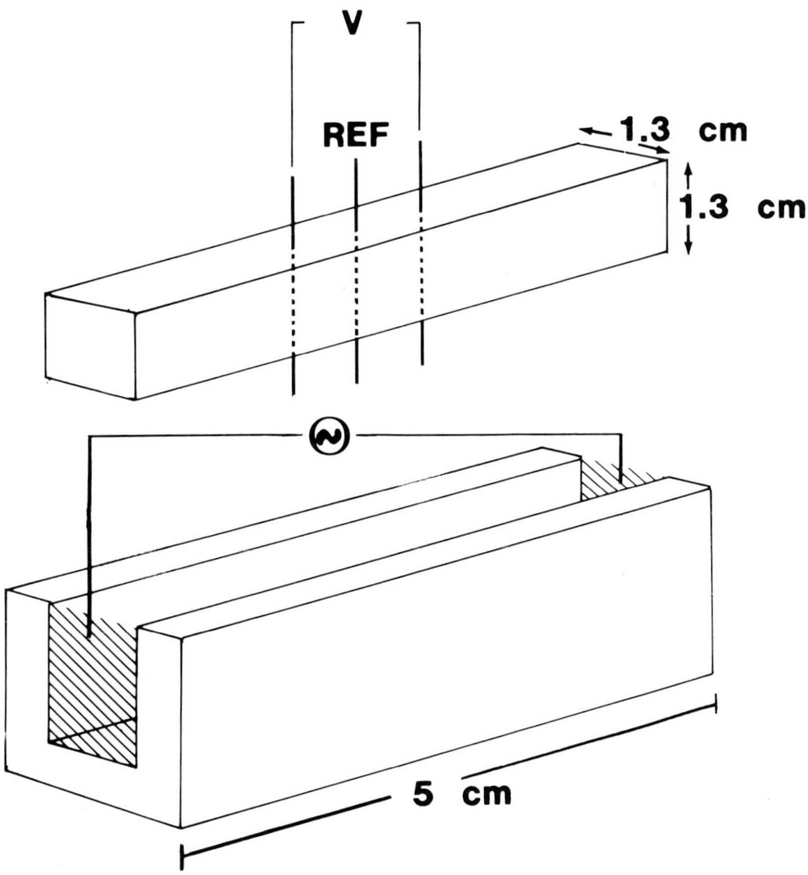

FIGURE 11B. Chamber for measurement of longitudinal spinal cord resistivity. Shaded region illustrates stainless steel current application electrodes. Voltage (V) measured with Nichrome wire electrodes inserted into the tissue.[17]

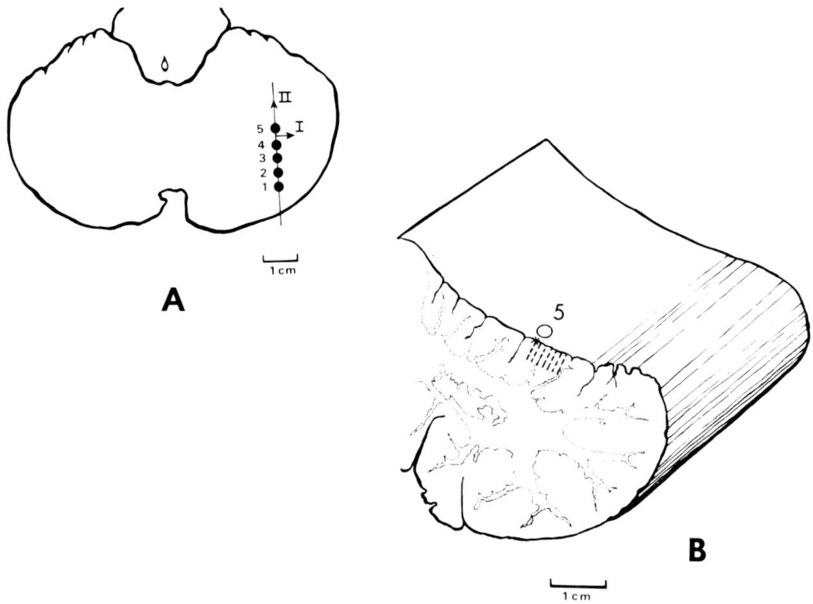

FIGURE 12. (A) Tracing of superior surface of unfixed human cadaver cerebellum showing location of electrodes for current density measurements. The arrows indicate axial and lateral directions in which the probe was moved for current density determinations (Tables 2 and 3). (B) Drawing of coronal block of unfixed human cadaver cerebellum to show electrode tracks between electrodes 4 and 5 at progressively lateral locations for measurement of current densities at various locations below the cerebellar surface (Tables 2 and 3).[3]

Table 1
CURRENT DENSITY OF HUMAN CEREBELLUM (mA/cm^2)[3]

Depth (mm)	Distance lateral (mm) 0	1	2	3	4	5	6
0	2.32	1.72	0.88	0.29	0.21	0.092	0.057
1	1.16	1.00	0.53	0.32	0.29	0.089	0.054
2	0.38	0.40	0.22	0.15	0.088	0.049	0.052
3	0.24	0.21	0.14	0.11	0.054	0.044	0.043
4	0.098	0.11	0.079	0.088	0.052	0.031	0.029
5	0.074	0.059	0.044	0.044	0.031	0.028	

Note: Average of four sets of data, two from each hemisphere, obtained at indicated distances from an electrode in a five-in-line array. (SDs were less than 30% in Tables 1 and 2.)

tracks for current measurements transverse to the axis of the five-in-line electrode set. The observations (Table 1) indicate that the current density is reduced to approximately 10% of the surface value at depths of 3 to 4 mm, and the surface current density is reduced by approximately the same amount at a corresponding distance laterally. Readings at depths greater than 5 mm were substantially less than those shown. Integration of the current values from Table 1 indicates that approximately 76% of the current passing from electrode 5 to electrode 4 was accounted for. The current density measured along the axis of the array was less at the edge of the array and was greatly decreased 2 to 3 mm below the surface.

In the monkey cerebellum, the current density decreased in depth to approximately the same degree as in the human (Table 2). The resistivity of the human tissue ranged from 500

Table 2
CURRENT DENSITY OF MONKEY CEREBELLUM (mA/cm^2)[3]

	Distance lateral (mm)					
Depth (mm)	**0**	**1**	**3**	**4**	**5**	**6**
0	5.2	4.18	2.07	1.07	1.05	0.63
1	1.84	1.96	0.77	0.61	0.43	0.36
2	1.02	0.56	0.50	0.38	0.38	0.20
3	0.40	0.32	0.23	0.17	0.30	0.15
4	0.18	0.15	0.20	0.13	0.071	0.076
5	0.10	0.081	0.14	0.080	0.057	0.046
6	0.039	0.049	0.069	0.053	0.032	0.026

Note: Average of two sets of data obtained at indicated distances from an active electrode.

to 1000 Ω-cm, with an average of 780 Ω-cm, and for the monkey if ranged from 500 to 800 Ω-cm, with an average of 630 Ω-cm.

To improve the effectiveness of cerebellar stimulation systems, additional studies were conducted to determine the current distribution with various electrode configurations.[14,15] An array of 17 electrodes was placed on the superior surface of the chimpanzee cerebellum, and a ten-electrode array was attached to the posterior surface (Figure 13).

The symmetry of the five-electrode array described above resulted in current flow mostly parallel to the axis of the array at points equidistant from positive and negative discs. The 17- and 10-disc arrays lacked symmetry and it was not possible to predict current direction in the cerebellum. Consequently each point required two current density determinations which were combined to yield the resultant current density.

To determine the magnitude and direction of the current density component in the x-y plane, the probe was inserted in the z direction and rotated to pick up maximum voltage and then loaded as usual[2,3,20] to find the magnitude of J. The direction of the measured J is normal to the probe surface. The probe was then removed and reinserted with its axis in the x-y plane and faces parallel to the plane. Loading the probe permitted measurement of the current density component in the z direction which was then added vectorially to the component from the x-y plane to yield the total current density. When the magnitude and/or direction of current density varies throughout the tissue, careful stereotactic placement of the probe is necessary to ensure that the two measurements are made at the same point. When all 27 electrodes on the posterior and superior cerebellar surfaces were connected together and current was passed to a distal point, the current flow was essentially perpendicular to the brain stem axis in a plane midway between the superior and posterior cerebellar surfaces (Figure 14A). The current from the distal electrode was directed radially inward towards the electrodes on the cerebellum. The current density in this plane ranged from 0.26 mA/cm^2 in the region of the superior-posterior edge of the cerebellum to 0.04 mA/cm^2 near the anterior surface. The currents in a perpendicular plane (Figure 14B) increased with proximity to the electrode arrays. These currents are perpendicular to the surface of the cerebellum. The calculated average current density at each electrode face was 2.5 mA/cm^2. The currents that reach the deep structures of the cerebellum are approximately one tenth of the surface values. The resistivity was approximately 300 Ω-cm near the surface and varied between 300 and 1200 Ω-cm elsewhere.

When the 17 electrodes on the superior surface were connected to 1 pole, and the 10 posterior electrodes were connected to the opposite polarity, a higher current density was

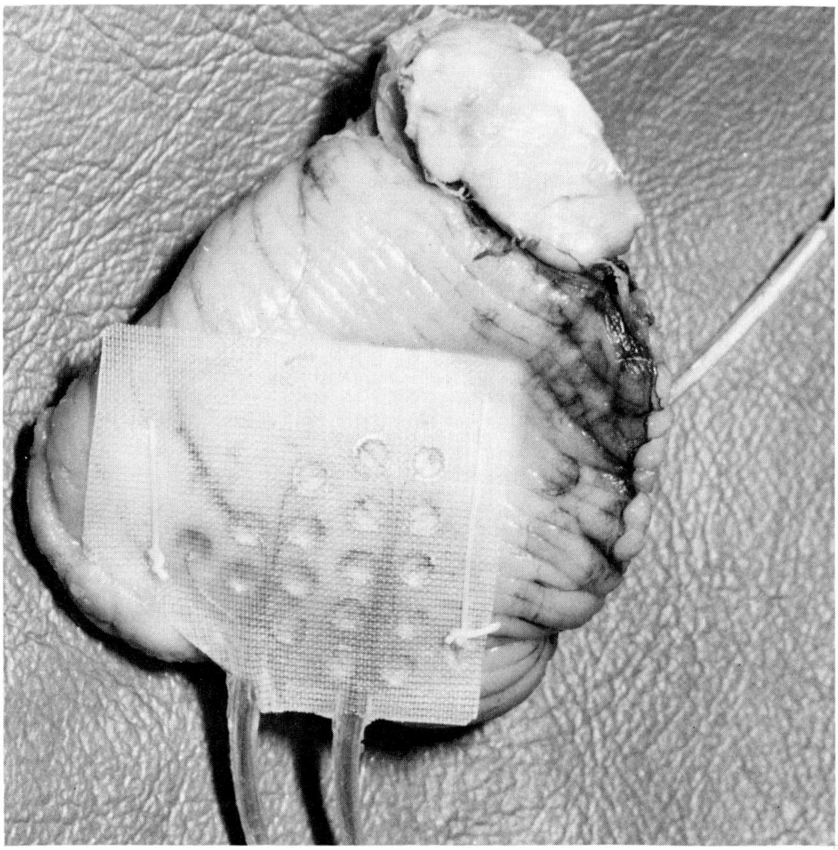

FIGURE 13. (A) Superior surface of chimpanzee cerebellum showing approximate location of electrode implant system. This system was attached to the cerebellum in preparation for the current density plots.

measured beneath the electrodes in the plane midway between the superior and posterior electrode sets (Figure 15). Substantially lower currents were measured near the lateral and medial surfaces. The majority of the current passes between the electrode sets in a direction parallel to the brain stem. While some spreading occurs with this array, it is substantially less than the previous connection (Figure 14).

Measurements in a perpendicular plane show an increase in current density near the electrodes (Figure 15b). While the calculated current at the surface of the superior electrode array is 4 mA/cm^2, the current densities decrease rapidly away from the electrode surface. In this plot, the currents are essentially parallel to the axis of the brain stem.

IV. EXTERNALLY APPLIED CURRENTS

The loaded probe technique works well in the evaluation of a current density secondary to spinal and cerebellar stimulation. However, with large dimensions for the tissue to be measured or the applied current electrodes, the four-electrode method is often preferable. Although the measurements discussed in this section were performed in conjunction with studies of electrical injury,[16,17] they are illustrative of the method and may yield insight to

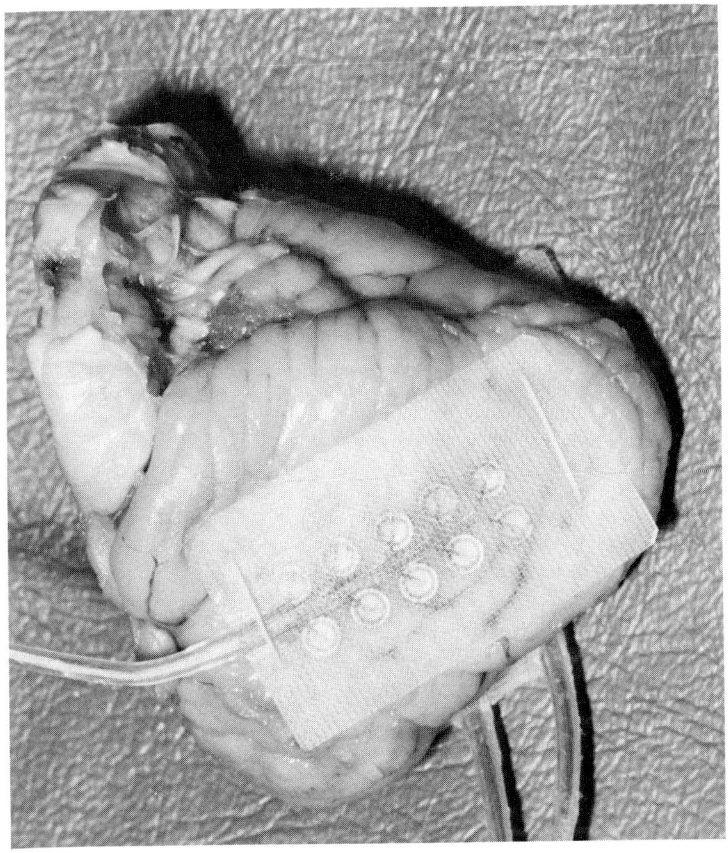

FIGURE 13B. Electrode array on posterior surface of chimpanzee cerebellum.[15]

the distribution of externally applied currents as used in transcutaneous neural stimulation.

The current density (J) was calculated by dividing measured electrical field intensity (E) by tissue resistivity (ρ).

$$J = E/\rho \tag{6}$$

To measure field intensity during application of 60-Hz limb-to-limb currents four-in-line electrode arrays (Figure 16) were inserted into the tissues. Voltage recorded between the inner two electrodes, $V_{2,3}$, was divided by electrode spacing (a) to give field intensity (Equation F, Table 3).

The four-in-line arrays were also used to measure tissue resistivities while the limb-to-limb currents were turned off. Resistivity in Ω-cm of a homogeneous isotropic media with the electrode on the surface is

$$\rho = \frac{2\pi a V_{2,3}}{I} \tag{7}$$

where I is the current applied to the outer two electrodes, and (a) and V are as defined above (Equations 6 and F). For electrodes within the tissue, tissue resistivity is

$$\rho = \frac{4\pi a V_{2,3}}{I} \tag{8}$$

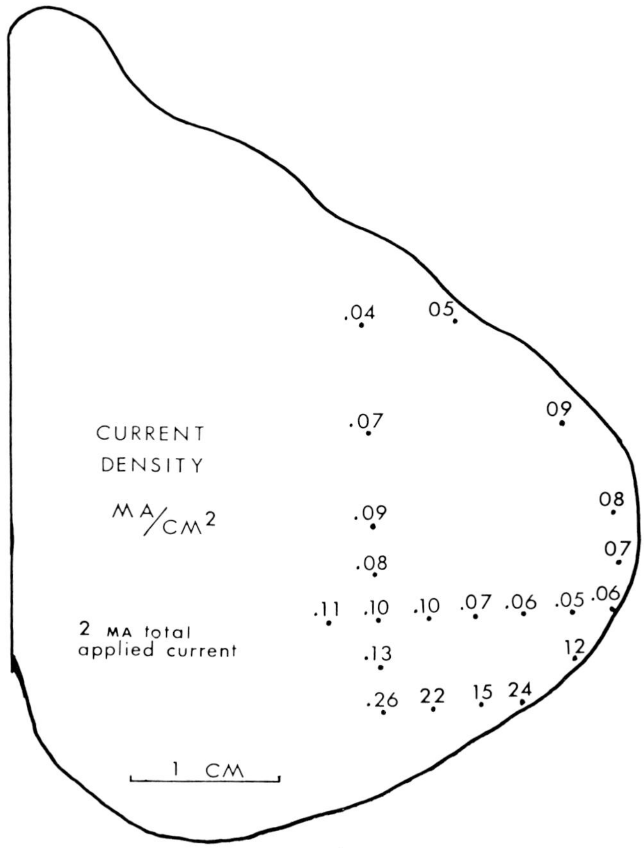

A

FIGURE 14. (A) Current density distribution in chimpanzee cerebellum with current applied to all cerebellar electrodes connected to one pole of a sinusoidal generator. A distal electrode provided the return path for current. Current density is essentially perpendicular to brain stem axis in the plane located midway between electrode sets. (B) Measurements taken at points in plane parallel to the brain stem axis. Principal current density component lies in plane of the paper and is parallel to the brain stem in regions near the electrodes and perpendicular to the brain stem midway between electrodes.[15]

It has been shown[11] that this equation is valid when thickness of the tissue layer is at least twice the electrode spacing. For resistivity measurements, currents were applied to the outer two electrodes to obtain the same voltage readings ($V_{2,3}$) as recorded during limb-to-limb current application.

Three different electrode array sizes were used to accommodate the different tissues in hogs. The smallest electrodes (1.8-mm spacing) measured resistivity of bone marrow and bone cortex (outer region of bone), skin, and leg muscle. The medium-sized (6-mm spacing) systems were used to measure currents flowing in lung, heart muscle, back muscle, liver, kidney, and fat. The largest system (24-mm spacing) was used in the intestine. When electrodes required surgical placement, the overlying tissues were sutured into their original positions. Resistivity and current density measurements of bone, bone marrow, tibial nerve, fat, muscle, and artery were made in the hindlimb, 7 cm above the distal end of the tibia (Figure 17).

Muscle measurements — Measurement techniques which take into account the anisotropicity of tissue are illustrated in Table 3. Resistivity was measured transverse and lon-

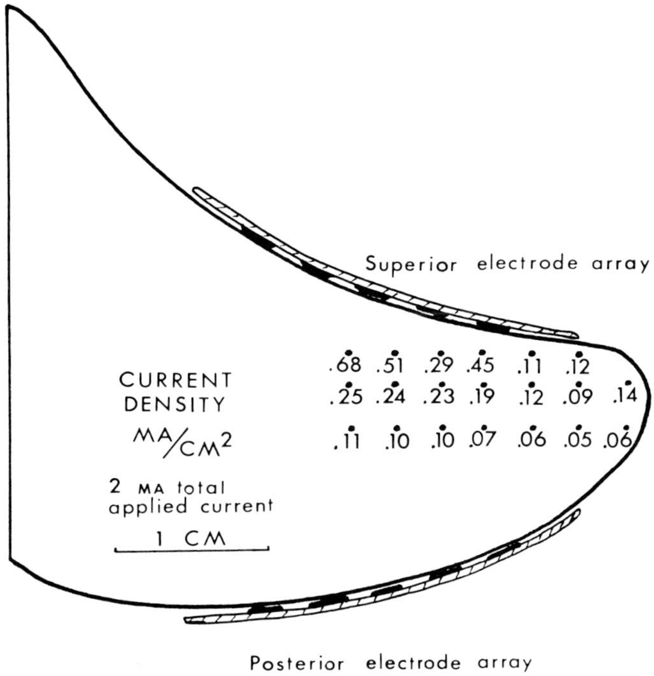

FIGURE 14B.

gitudinal to the muscle fibers in the leg. The four-electrode set was placed longitudinally on the fibers and transverse resistivity was found using Equation A, Table 3. This calculation assumes that transverse resistivity is the same into and across the muscle. Longitudinal resistivity is found by placing the electrode set transverse to the fibers. Voltage measured transverse to the fibers is combined with the longitudinal measurement above to give the longitudinal resistivity using Equation B, Table 3.[11] Current density in the longitudinal and transverse directions can then be calculated from Equations G and H, Table 3.

Nerve and vessel — The four-in-line arrays could not be used in the small-diameter tibial nerve and artery. Instead, two 0.13-mm diameter Teflon®-coated silver wires were inserted 5 mm apart through the nerve and artery. The wires were bare only inside the tissue for field measurements during application of the 60-Hz potentials. At the end of the study, the tibial nerve was cut 2 cm above the measurement site. Three to five cm of the nerve were raised above the skin surface and a 20- to 100-μA. 60-Hz current was passed from the electrode on the limb to a silver electrode plate on the cut nerve end. The average cross section of nerve was calculated from its measured length and the volume of saline it displaced. The resistance, resistivity, and current density were then found using Equations C, D, and I in Table 3.

With limb-to-limb contact using gelled electrodes, approximately 15 V were required from 30-mA current flow, 40 V for 100 mA, and 415 V for 1 A. The resistivity of each of the tissues in the hindlimb, 7 cm above the distal end of the tibia, is shown in Table 4. The phase angle between the current and voltage was less than 5° for all tissue measurements. The current densities were approximately inversely proportional to the resistivities of the respective tissues. A 5 to 10% difference in current density was observed with changing position of the hindlimbs of the hog. The approximate percentages of the cross-sectional area of the tissue shown in Figure 17 are 20% skin, 37% muscle, 0.4% blood vessel, 5% tendon, 3% bone marrow, 5% bone marrow, 29% fat, and 0.3% nerve.

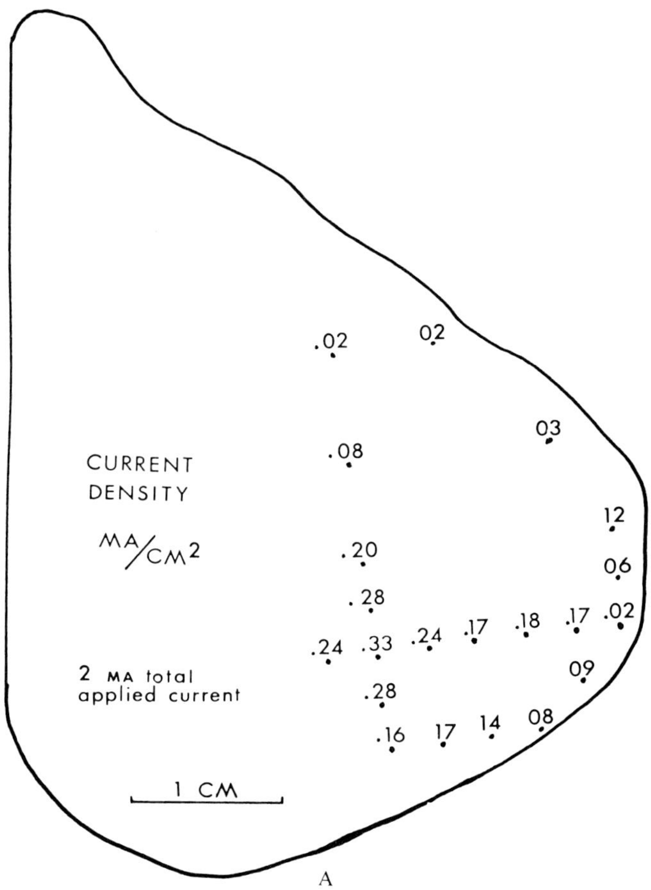

FIGURE 15. Current density distribution in chimpanzee cerebellum secondary to current application with all superior electrodes connected to one pole of sinusoidal generator and all posterior electrodes connected to the other pole. (A) Measurements were made in horizontal plane midway between superior and posterior electrode sets. Current flow is essentially perpendicular to the plane. (B) Measurements taken at points in a plane parallel to the brain stem axis. Principal density component is parallel to the brain stem axis.[15]

V. DISCUSSION

Current density measurement in biological systems continues to be a challenge. Although a variety of techniques have been employed, none are universally used. Often, they must be tailored to a specific situation.

The loaded-probe technique works well for measuring distribution of externally applied 1000-Hz currents in the CNS. The measurement yields current density directly without the necessity of injecting additional current for a resistivity measurement, as would be required by the three- and a four-electrode methods. Besides being inconvenient, additional current may excite some neurons and alter measured resistivity.

Loaded probe theory assumes that the medium in which measurements are made is electrically linear, homogeneous, and isotropic. The tissue should be linear over the range of field intensities encountered in the process of loading the probe. Usually this range will vary by less than a factor of 2. With respect to homogeneity, the probe will measure accurately if it is placed at least one or two times its size away from the boundary between tissues of differing resistivities.

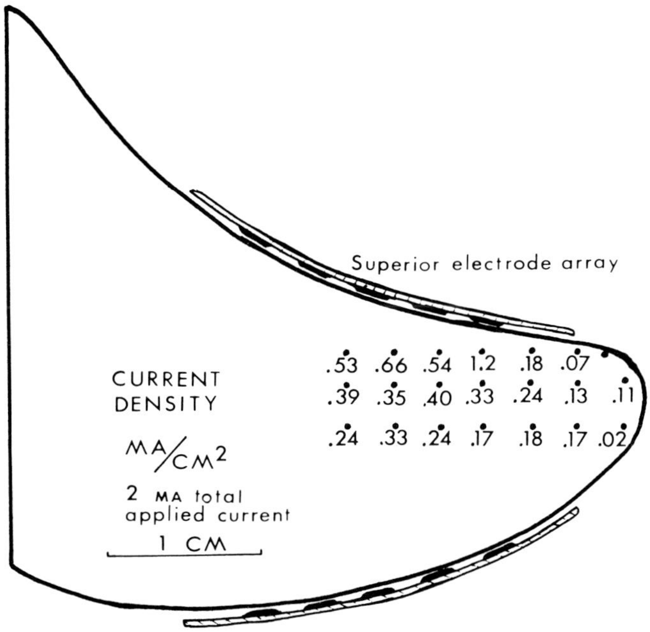

FIGURE 15B.

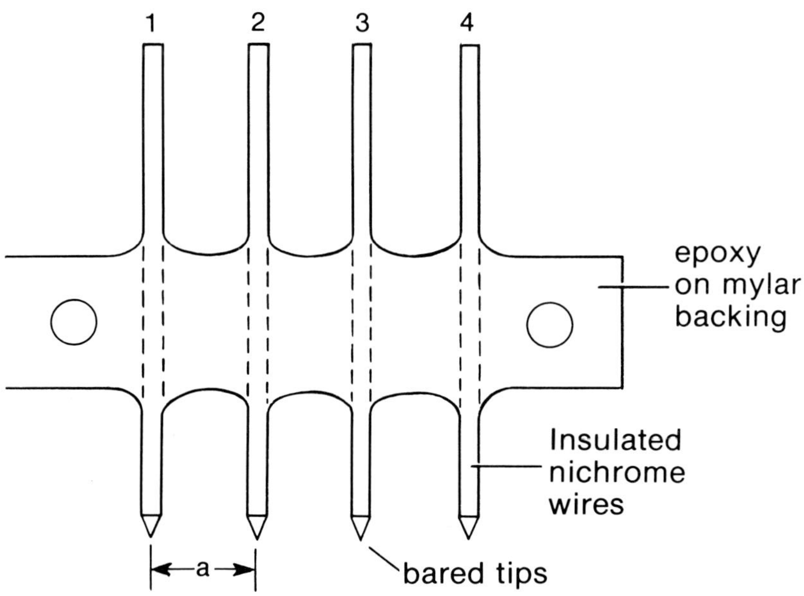

FIGURE 16. Four-in-line electrode current density measuring system. The spacing, a, was 1.8, 6, or 24 mm. Current is applied between leads 1 and 4, with voltage measurements between leads 2 and 3 (see Table 3).[17]

If the probe is used in anisotropic media, the probe factor will theoretically be different from that in isotropic media. Thus, far, no one has demonstrated how this difference can be quantified.

Table 3
RESISTIVITY MEASUREMENTS IN VARIOUS TISSUES[17]

Tissue	Resistivity $\rho(\Omega\text{-cm})$	Current density $J\left(\frac{mA}{cm^2}\right)$
1.8-mm spaced electrode Skin, bone marrow, bone cortex, leg muscle 6-mm spaced electrode Abdominal wall, heart muscle, back muscle, liver, kidney, fat, lung 24-mm spaced electrode Ventral intestine, dorsal intestine	4 Electrodes 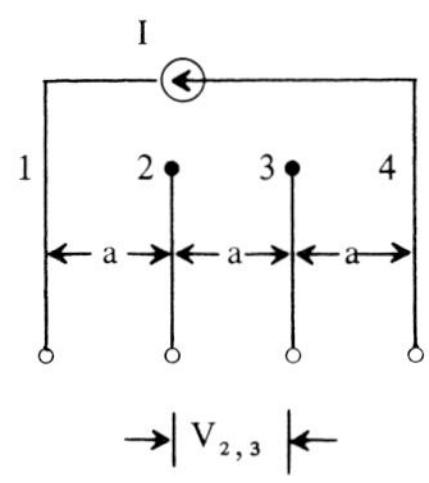 Current application electrodes 1 and 4 Voltage measurement The transverse resistivity is $\rho_T = \frac{2\pi a V_{2,3}}{I}$ (A) for a surface measurement where $V_{2,3}$ is measured in the longitudinal direction (V_L) of tissue fibers The relationship between the longitudinal (ρ_L) and transverse resistivity (ρ_T) is $\rho_L = \left(V_T/V_L\right)^2 \rho_T$ (B) where V_T is $V_{2,3}$ in the transverse direction to the tissue fibers	The electric field measured between 2,3 is $E = \frac{V_{2,3}}{a}$ (F) due to 60 Hz applied between the limbs; the current density in the longitudinal direction is $J_L = E_L/\rho_L$ (G) where E_L is the electric field in the longitudinal direction The current density in the transverse direction is $J_T = E_T/\rho_T$ (H) where E_T is the electric field in the transverse direction
Peripheral nerve Arterial measurements, the same as nerve except the blood was drawn and ρ measured in an impedance chamber	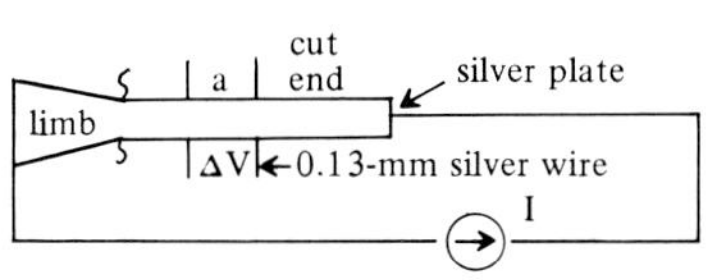 The resistance is $R = \frac{\Delta V}{I}\ \Omega$ (C) The resistivity is $\rho = \frac{RA}{a}\ \Omega\text{-cm}$ (D) where A = cross-sectional area and a = electrode spacing I = applied current	The electric field measured with 60 Hz applied between the limbs $E = \Delta V/a$ (I) the current density is $J = E/\rho$ (J)
Spinal cord	See Figure 11 In vitro measurement for resistivity $\rho = \frac{EA}{I}$ (E) where A is the cross section of the spinal cord contacting the current plates; Equation 6 is used for both longitudinal and transverse resistivity measurements	Current density is calculated the same as in the nerve

Another assumption of the loaded-probe technique is that the impedance encountered at the electrode-electrolyte interface is negligibly small compared to the load resistance. The electrode-electrolyte impedance varies inversely with the surface area of the probe. Since

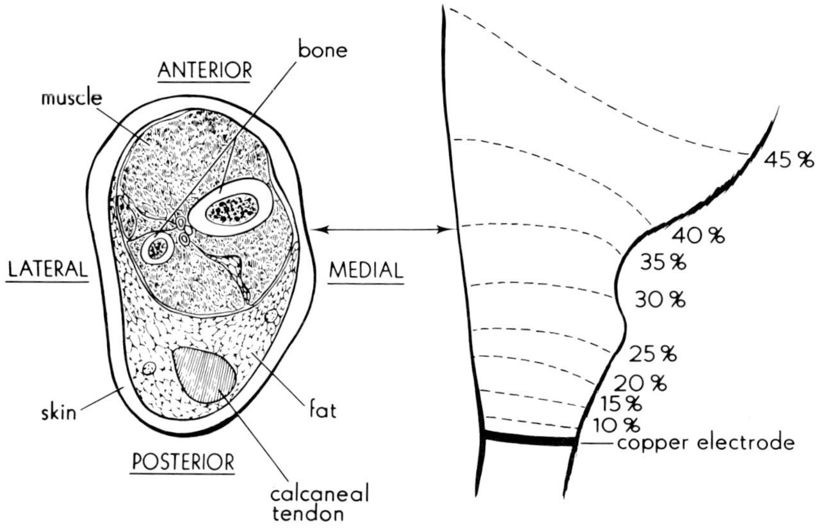

FIGURE 17. Experimental setup for current density measurements with cross section where resistivities and current densities were evaluated. The lines on the limb (right) indicate percentages of the total applied voltage measured in the dermis from the copper input electrode to proximal regions of the animal.[17]

surface area is proportional to the probe size squared, a 0.3-mm probe would have about one tenth the surface area of a 1-mm probe and about ten times greater interface impedance. In contrast, the proper load resistance varies inversely with probe size (raised to the first power) and would be about 3.3 times greater for a 0.3-mm probe compared to a 1-mm probe. Because interface impedance increases more than proper load resistance, there is a practical limit as to how small a probe can be used and still provide accurate readings. Frequency of measured current and the medium resistivity also influences this limit since they affect interface impedance and size of the proper load resistance.

To reduce the electrode-electrolyte interface impedance, the probe was heavily coated with platinum black. This coating was delicate and easily rubbed off when the probe was inserted into solid tissue. In the soft tissue of brain and spinal cord, enough coating remained on the 0.3-mm probes to permit measurements at 1000 Hz. However, at 60 Hz the voltage under load was lower than it should have been, indicating a significant voltage drop across the interface impedance. At higher frequencies, impedance is lower and measurements are possible.

As an alternative to the loaded probe, a four-electrode array was used to measure current density at 60 Hz. The electrode-electrolyte impedance does not affect measurements over a wide range of frequencies because the input impedances of the voltage amplifiers are generally much higher than the interface impedance. Therefore, the voltage drop across the interface impedance is neglible. As with the loaded probe, resistivity and current density measured with the four-electrode method depends primarily on tissue closest to the electrodes. Tissue more than twice the electrode spacing away from the electrode array has little effect on the measurement. However, electrode arrays as small as the probe tend to bend either during placement in tissue or during muscular contraction due to applied current. This change in electrode spacing reduces measurement accuracy. Surface electrode arrays allow muscle tissue to contract without changes of electrode spacing.

Generally, both measurement techniques assume isotropic media. However, the four-electrode method can handle some anisotropic media if they are describable by three principal resistivities in mutually orthogonal directions. These directions must be known *a priori*. Also a separate measurement must be made in a different direction for each resistivity.

Table 4
TISSUE CURRENT DENSITY, J (in mA/cm^2), AND RESISTIVITY, P (in Ω-cm), IN THE LEG CROSS SECTION (FIGURE 2) VS. APPLIED CURRENT FROM HINDLIMB TO HINDLIMB (n, number of measurements)

		Artery		Nerve		Muscle			Fat		Bone marrow		Bone cortex	
Total current (mA)	Approximate applied voltage (V)	J (n = 7)	ρ	J (n = 8)	ρ	J (longitudinal, n = 11)	ρ (longitudinal)	ρ (transverse)	J (n = 5)	ρ	J (n = 4)	ρ	J (n = 4)	ρ
10	6	0.32/0.04	147/10	0.26/0.05	201/42	0.18/0.15	296/50	512/67	0.15/0.07	380/80	0.09/0.03	570/100	0.04/0.01	1900/250
100	45	3.4/0.48	152/9	3/0.85	191/50	2/0.48	295/59	483/70	1.4/0.70	340/100	1.2/0.20	525/105	0.30/0.05	1850/310
1000	415	35.9/2.7	155/15	27.1/7.2	200/51	19.5/8.2	290/50	650/120	14.9/4.2	365/98	8/1.7	500/110	2.9/0.7	1839/295

Note: Entries given are mean/SD.

With the four-electrode method, current density has been determined from field intensity measurements made during application of externally applied current and tissue resistivity measured without externally applied current. Separate measurements were necessary so that tissue resistivity could be found by using a small current at the same frequency as the larger externally applied current. Recently, however, resistivity has been found from a small 70-Hz signal during the application of external 60-Hz currents. With this technique, alterations in resistivity secondary to externally applied currents can be determined.

In summary, evaluation of current distribution has aided the development of spinal and cerebellar implants systems. The techniques discussed above may also help in evaluating new systems involving therapeutic application of electrical current.

ACKNOWLEDGMENT

This research was supported in part by the Veterans Administration Medical Center Research Grant #6000-02P and the NIH Research Grant #GM26956.

REFERENCES

1. **Coburn, B.,** Electrical stimulation of the spinal cord: two-dimensional finite element analysis with particular reference to epidural electrodes, *Med. Biol. Eng. Comput.*, 18, 573, 1980.
2. **Jarzembski, W.B.,** Current density measurement in living tissue, in *Neural Stimulation*, Vol. 1, Myklebust, J. B., Sances, A., Jr., Larson, S. J., and Cusick, J. F., Eds., CRC Press, Boca Raton, Fla., 1985, chap. 3.
3. **Larson, S. J., Sances, A. Jr., Cusick, J. F., Myklebust, J., Millar, E. A., Boehmer, R., Hemmy, D. C., Ackmann, J. J., and Swiontek, T. J.,** Cerebellar implant studies, *IEEE Trans. Biomed. Eng.*, 23(4), 319, 1976.
4. **Larson, S. J., Sances, A., Jr., Cusick, J. F., Meyer, G. A., and Swiontek, T.,** A comparison between anterior and posterior spinal implant systems, *Surg. Neurol.*, 4, 180, 1975.
Neurosurgery, (2), 212, 1977.
6. **Larson, S. J., Sances, A., Jr., Hemmy, D. C., Millar, E. A., and Walsh, P. R.,** Physiological and histological effects of cerebellar stimulation, *Appl. Neurophysiol.*, 40(2-4), 160, 1977/78.
7. **Larson, S. J., Sances, A. Jr., Riegel, D. H., Meyer, G. A., Dallmann, D. E., and Swiontek, T.,** Neurophysiological effects of dorsal column stimulation in man and monkey, *J. Neurosurg.*, 41(2), 217, 1974.
8. **Myklebust, J., Sances, A., Jr., Swiontek, T., Larson, S. J., and Maiman, D.,** Spinal cord stimulation studies, in *Proc. 6th Int. Symp. Electrostimulation*, Albena, Bulgaria, September 24, 1981, 216.
9. **Plonsey, R.,** Electrical field modeling, in *Neural Stimulation*, Vol. 1, Myklebust, J. B., Sances, A., Jr., Larson, S. J., and Cusick, J. F., Eds., CRC Press, Boca Raton, Fla., 1985, chap. 2.
10. **Ranck, J. B., Jr. and Bement, S. L.,** The specific impedance of the dorsal columns of cat: an anisotropic medium, *Exp. Neurol.*, 11, 451, 1965.
11. **Rush, S.,** Methods of measuring the resistivities of anisotropic conducting media in situ, *J. Res. Natl. Bur. Stand.*, 66c, 217, 1962.
12. **Rusinko, J. B., Walker, C. F., and Sepulveda, N. G.,** Finite element modeling of potentials within the human thoracic spinal cord due to applied electrical stimulation, *IEEE 1981 Frontiers of Engineering in Health Care*, IEEE, Piscataway, N. J., 1981, 76.
13. **Sances, A. Jr. and Larson, S. J.,** *Electroanesthesia: Biomedical and Biophysical Studies*, Academic Press, New York, 1975.
14. **Sances, A., Jr., Larson, S. J., Myklebust, J., Swiontek, T., Millar, E., Cusick, J. F., Hemmy, D. C., Jodat, R., and Ackmann, J. J.,** Studies of electrode configuration (Symposium: Neuroaugmentive Session on Chronic Cerebellar Stimulation), *Neurosurgery*, 1(2), 207, 1977.
15. **Sances, A., Jr., Larson, S. J., Myklebust, J., Swiontek, T., Millar, E. A., Cusick, J. F., Hemmy, D. C., Jodat, R., and Ackmann, J. J.,** Evaluation of electrode configurations in cerebellar implants, *Appl. Neurophysiol.*, 40(2—4), 141, 1977/78.
16. **Sances, A., Jr., Myklebust, J. B., Larson, S. J., Darin, J. C., Swiontek, T., Prieto, T., Chilbert, M., and Cusick, J. F.,** Experimental electrical injury studies, *J. Trauma*, 21(8), 589, 1981.

17. **Sances, A., Jr., Myklebust, J. B., Szablya, J. F., Swiontek, T. J., Larson, S. J., Chilbert, M., Prieto, T., Cusick, J. F., Maiman, D., and Pintar, K.,** Current pathways in high voltage injuries, *IEEE Trans. Biomed. Eng.*, 30(2), 118, 1983.
18. **Sances, A., Jr., Swiontek, T. J., Larson, S. J., Cusick, J. F., Meyer, G. A., Millar, E. A., Hemmy, D. C., and Myklebust, J.,** Innovations in neurologic implant systems, *Med. Instrum.*, 9(5), 213, 1975.
19. **Sepulveda, N. G. and Walker, C. F.,** Finite element modeling of potentials within the human brain due to applied electrical stimulation, in *2nd IEEE-EMBS Conf.*, IEEE, Piscataway, N.J., 1980.
20. **Swiontek, T. J., Sances, A., Jr., Larson, S. J., Ackmann, J. J., Cusick, J. F., Meyer, G. A., and Millar, E. A.,** Spinal cord implant studies, *IEEE Trans. Biomed. Eng.*, 23(4), 307, 1976.
21. **Swiontek, T., Sances, A., Jr., Larson, S. J., and Cusick, J. F.,** Current density and impedance measurements in nervous tissue. Proc. 5th International Symposium: Impedance Methods for Brain Circulation Investigators, *Neurologija (Zagreb)*, 28(1—4), 147, 1980.
22. **Swiontek, T., Maiman, D., Sances, A., Jr., Myklebust, J., Larson, S. J., and Hemmy, D.,** Effect of electrical current on temperature and pH in cerebellum and spinal cord, *Surg. Neurol.*, 14(5), 365, 1980.

Chapter 5

SOMATOSENSORY EVOKED POTENTIALS IN EXPERIMENTAL NEURAL STIMULATION

J. Myklebust and J. Cusick

TABLE OF CONTENTS

I. INTRODUCTION

The somatosensory evoked potential (SSEP) is extensively used in the diagnosis of peripheral nerve disorders,[6,15,26] spinal cord pathology,[17,18,72] and cerebral function.[28,64] It is useful for monitoring patients in intensive care[1-3,35,36,47] and the evaluation of coma and brain death.[4,22] Applications have been reported for the serial evaluation of stroke and degenerative diseases such as multiple sclerosis[42,46,60] and cervical spondylotic myelopathy.[38] It is increasingly applied to intraoperative monitoring and for the localization of CNS structures.[29,43,45,67] The SSEP is an ideal tool for the evaluation of particular neural systems and structures in the experimental animal. With implanted electrodes, the rapidly conducting specific afferent, nonspecific afferent, and efferent systems can be monitored.

II. CHARACTERISTICS OF THE SSEP

A. Recording Locations and Electrodes

Although many animals have been used in studies of evoked potentials, neuroanatomical differences make nonhuman primates the best choice for evaluation of neural system effects. The studies reported below were done in the stumptail (*Macaca arctoides*) and rhesus (*M. mulatta*) monkeys. For stimulating and recording from the surface of the brain or spinal cord, 3-in-line, or 5-in-line 2-mm-diameter platinum discs, 0.125-mm-thick, spaced 4 mm on center, and embedded in Dacron®-reinforced silastic are used. These electrodes are placed bilaterally into the epidural space over the sensorimotor cortices (SMC) and are positioned through small openings in the interlaminal ligaments into the dorsal epidural space over the cauda equina (CE), conus medullaris, and upper thoracic or lower cervical spinal levels. Bipolar Nichrome electrodes, 0.25-mm-diameter, separated 0.3 mm on center and insulated except at the tips, are used for stimulating and recording within the brain. Typically, electrodes are stereotaxically implanted into thalamic nucleus ventralis postero lateralis (VPL) or in the thalamic intralaminar nuclei, centre median (CM), or parafasicularis (PF).

Teflon®-coated 0.0125-mm-diameter stainless steel leads are brought out from the electrodes to Amphenol connectors. The leads and connector from cerebral electrodes are incorporated into a methylmethacrylate block fixed to the calvarium. Leads from spinal electrodes are led out through the skin. When the electrodes are not being used, the connectors are placed in silastic envelopes and positioned in individual s.c. pockets.

Although it is sometimes necessary to surgically expose peripheral nerves, it is usually sufficient to stimulate the nerves transcutaneously using needle electrodes. Frequently used peripheral stimulation sites include the median nerve at the wrist, the common peroneal nerve at the fibular head, the sural and posterior tibial nerves at the ankle, and the femoral nerve in the femoral triangle.

With peripheral nerve stimulation, afferent systems can be evaluated at the level of peripheral nerve (cauda equina), spinal dorsal root entry zone (conus medullaris or lower cervical spinal cord), at the upper spinal cord, and at thalamic and cortical cerebral levels. By stimulating implanted electrodes, transmission through various sections of the CNS can be observed. With stimulation of the sensorimotor cortex, the integrity of primary efferent pathways can be followed by recording from the spinal cord.

For most of these studies, recordings are made between relatively closely spaced electrodes (bipolar recordings). Closely spaced electrodes reflect local activity, while widely spaced electrodes are likely to record composites of signals from widely spaced neural generators. The choice of which recording system to use depends upon the purpose of the recording. In clinical applications it may be desirable to detect several generators in one averaged response.[41] However, if it is desired to study a particular neural center (e.g., specific somatosensory cortex), it is usually preferable to exclude contributions from other sites.

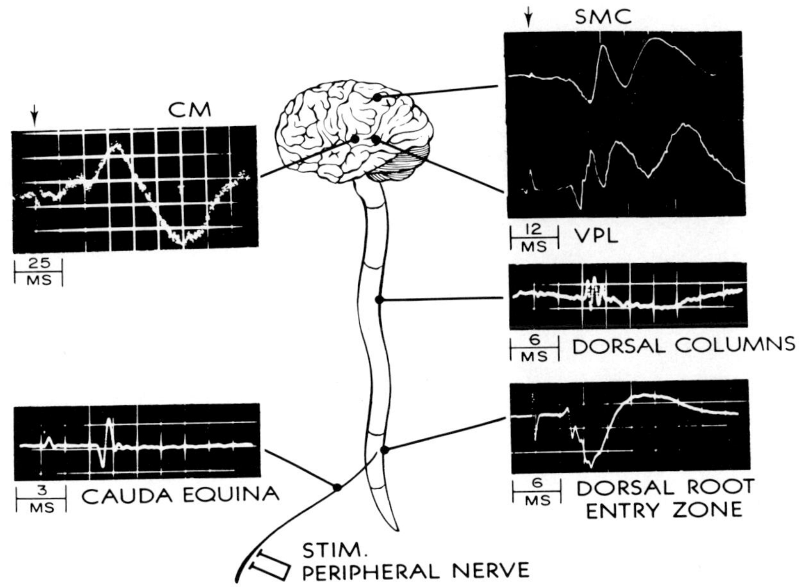

FIGURE 1. Typical recordings obtained from the stumptail (*M. arctoides*) monkey, with stimulation applied to the common peroneal nerve at the fibular head. Responses from cauda equina, dorsal root entry zone, dorsal columns, sensorimotor cortex (SMC) and thalamic nuclei ventralis postero lateralis (VPL), and centre median (CM). Stimulus at arrow.

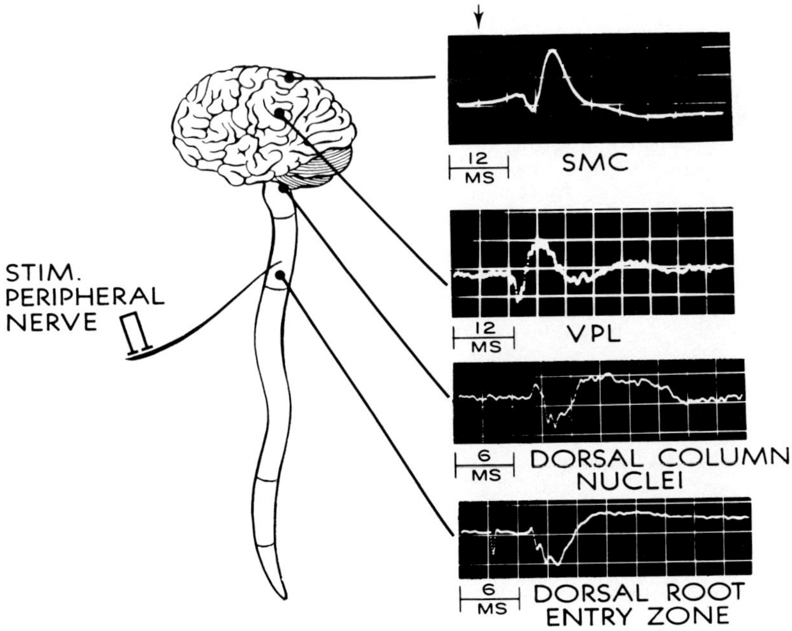

FIGURE 2. Recordings obtained from lower cervical spinal cord (dorsal root entry zone), cervico-medullary junction (dorsal column nuclei), thalamic nucleus VPL, and sensorimotor cortex secondary to stimulation of the median nerve of the monkey. Stimulus at arrow.

B. Typical Recordings

Evoked potentials recorded from various regions of the CNS secondary to peripheral nerve stimulation are shown in Figures 1 to 3. A complex waveform, typical of neural tracts, is

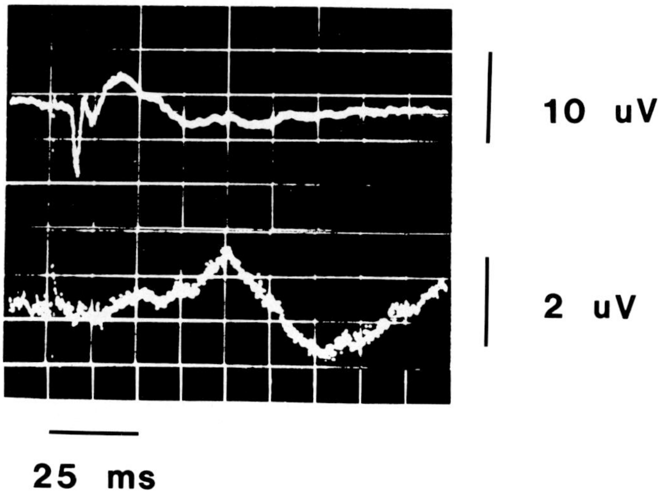

FIGURE 3. Comparison of responses from specific relay nuclei (top, thalamic nucleus VPL) and nonspecific nucleus (bottom, intralaminar thalamic nucleus centre median) with peroneal nerve stimulation in the monkey. Stimulus after first 10% of sweep.

observed at cauda equina, thoracic spinal cord, medial lemniscus, and in thalamocortical fibers (Figure 1). The relatively slow potentials recorded from the spinal cord at levels where the dorsal roots enter the cord, over the dorsal column nuclei in the medulla, in nucleus ventralis postero lateralis of the thalamus, and over the sensorimotor cortex are typical of synaptic regions (Figure 2).

Potentials recorded from nonspecific afferent systems as typified by the thalamic intralaminar nuclei, demonstrate longer latencies and monophasic waveforms in comparison to responses from specific systems (Figure 3). Furthermore, the nonspecific responses are usually elicited by either ipsilateral or contralateral stimulation.

C. Neural Substrate

From clinical studies, the cortical SSEP has been suggested to be a reflection of the dorsal column system. [31,37,44] However, conflicting experimental results have been reported.[76] To determine the pathways implicated in the transmission of the SSEP, a series of ablation experiments was performed.[16,17]

When the dorsal columns were selectively resected, stimulation of distal peripheral nerve or dorsal column showed that responses cephaled to the lesion in the spinal cord, VPL, and in SMC were eliminated. In contrast, responses in the intralaminar nuclei of the thalamus remained (Figure 4).

Dorsolateral column resection and anterior column resection had no significant effect on the SSEP recorded in the spinal cord, VPL, or SMC. However, the response in the intralaminar nuclei was obliterated with anterior column resection.

With isolated preservation of the dorsal columns, the SSEP in the spinal cord, VPL, and SMC was unaltered. However, the response in the intralaminar nuclei was obliterated (Figure 5).

Consequently, responses recorded from the spinal cord rostral to the dorsal root entry zone, in nucleus VPL of the thalamus, and over sensorimotor cortex with peripheral nerve or dorsal spinal cord stimulation primarily reflect activity in the dorsal column-medial lemniscal system. Cortical responses due to other systems are not directly observed with the stimulus parameters normally used. This is probably due to the slower conduction velocity

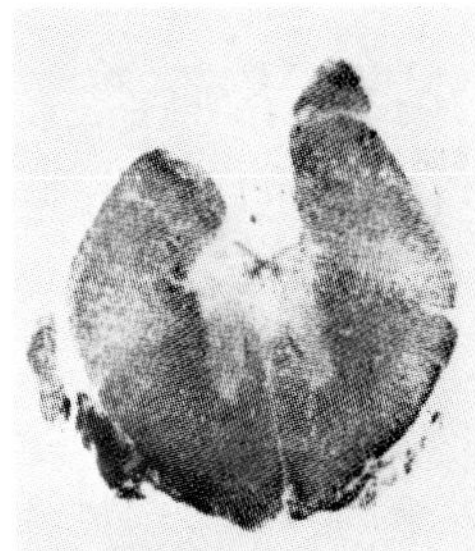

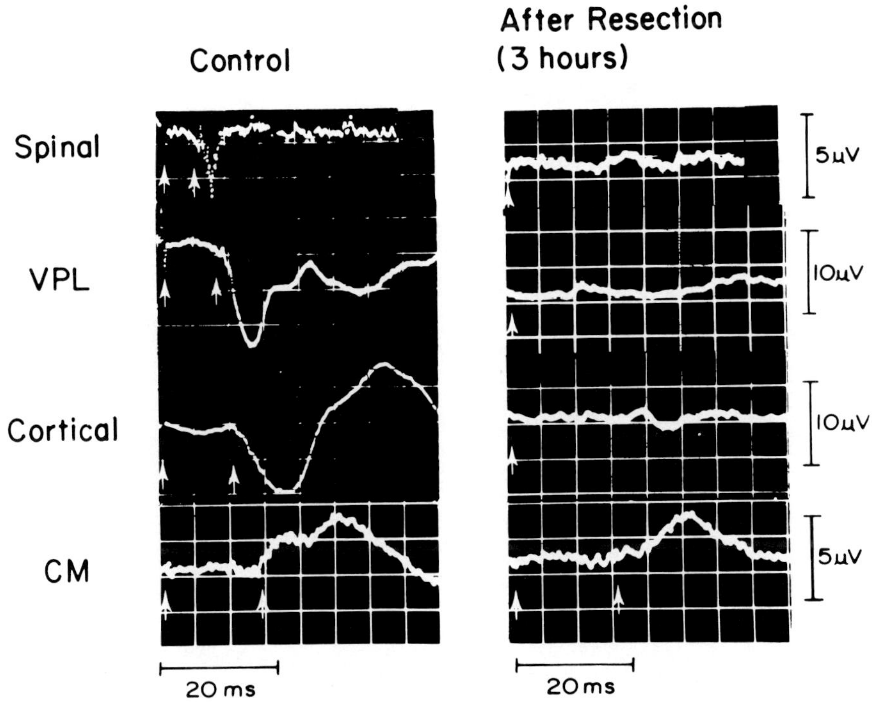

FIGURE 4. Effect of dorsal column ablation upon responses over dorsal spinal cord (spinal), sensorimotor cortex (cortical), and thalamic nuclei. VPL CM with peroneal nerve stimulation. Stimulus at first arrows.[17]

and nonspecific nature of these other systems, which results in long-latency, low-amplitude, monophasic potentials. Further, the stimulus parameters routinely used do not fully activate the small fibers which comprise these systems. However, the nonspecific systems may modulate the after-potentials of the SSEP by means of thalamothalamic relationships.

Responses secondary to activity in the nonspecific systems, probably transmitted through ventral spinal columns, can be recorded in the thalamic intralaminar nuclei. Although the connections of these nuclei are not clear, clinical data suggest that they are involved in the perception of pain.[40,63,87]

The nonspecific thalamic nuclei (ventral anterior, midline, and intralaminar nuclei) are characterized by diffuse projections to the cortex which are often collateral or sustaining. On stimulation, long latency recruiting responses are produced over widespread cortical regions. The recruiting response increases with stimulus frequency in the range of 6 to 12 Hz.[43] The recruiting response is a useful feature in the electrophysiologic verification of electrode position in the intralaminar nuclei (Figure 6).

In their pioneering study of electrophysiology of thalamocortical connections, Dempsey and Morison[24,25,56] found that the recruiting response to stimulation of the thalamic intra-

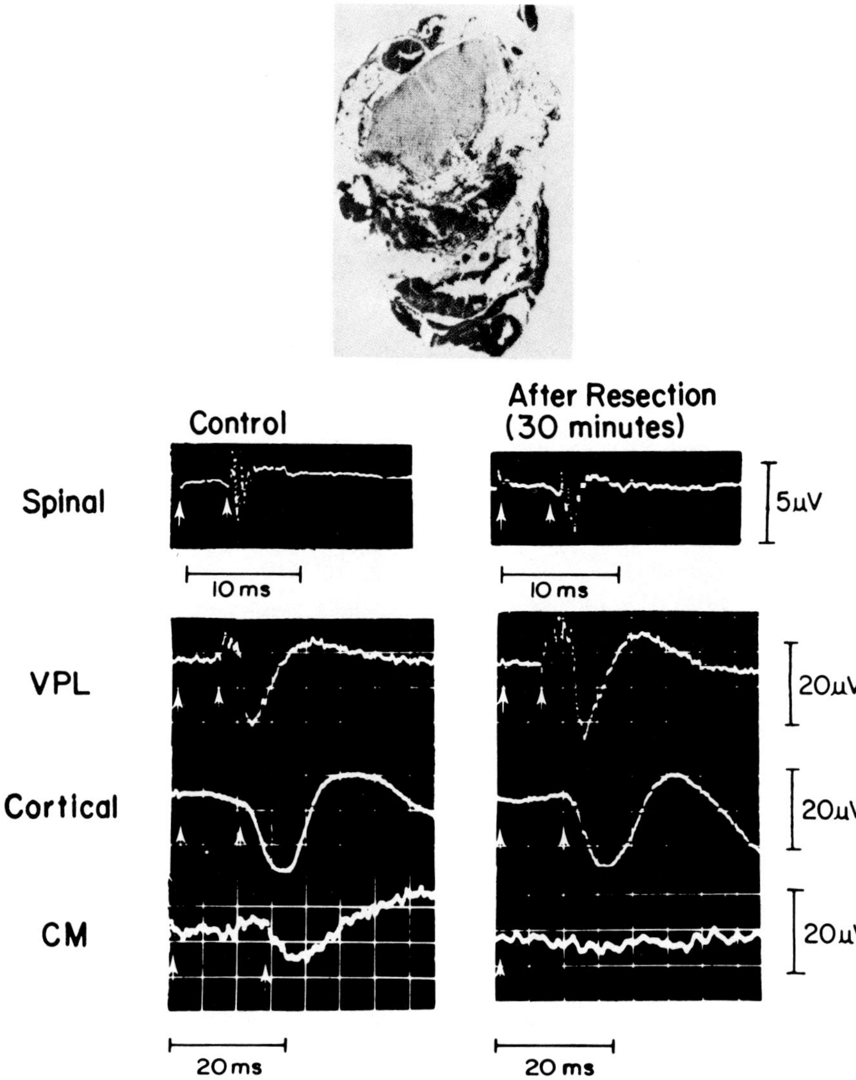

FIGURE 5. Effects of elimination of entire spinal segment except dorsal columns upon responses recorded from dorsal spinal cord (spinal), sensorimotor cortex (cortical), and thalamic nuclei VPL and CM with peroneal nerve stimulation. Stimulus at first arrow.[17]

laminar nuclei is conducted by a diffuse, nonspecific projection to cortex.[56] The primary response to peripheral stimulation was found to be unaffected by the recruiting potentials as well as 8- to 12-Hz EEG potentials which occur spontaneously in cortex. The recruiting potentials and the spontaneous 8- to 12-Hz potentials were found to be mutually exclusive.[24,25] These observations led to the conclusion that the recruiting potentials are conducted by the same thalamocortical projection which produces the spontaneous 8- to 12-Hz rhythmic EEG potentials. The nonspecific nuclei apparently regulate the excitability of the cortex through thalamothalamic and thalamocortical connections[61,65] and are probably involved in the ascending reticular activating system.

The potential recorded from the dorsal root entry zone for the stimulated peripheral nerve has a characteristic waveform reflecting the segmental spinal activity (Figure 2). The initial notches reflect the activity of the primary afferent fibers in the dorsal roots and dorsal column.[5,23] The long-duration negative portion of the waveform represents the synaptic

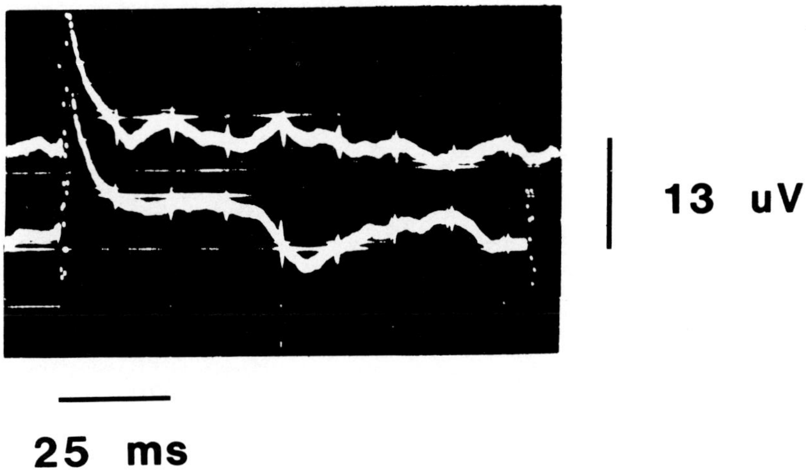

FIGURE 6. Illustration of recruiting response recorded over sensorimotor cortex with stimulation of intralaminar thalamic nuclei. Upper trace is with 7-Hz and lower trace is with 9-Hz stimulation frequency. Stimulus is after first 10% of sweep.

activity of the internuclial and relay cells of the dorsal horn, and the slow wave probably results from synaptically mediated depolarization of the terminals of incoming fibers.[27,88] Over the dorsal spinal cord at levels rostral to the dorsal root entry zone, only a polyphasic tract response is seen. This response is temporally related to the initial notches of the segmental response, follows at high frequency, and is relatively insensitive to ischemia. This response represents the activity propagated in the dorsal columns by primary afferent fibers.

D. Spinal Stimulation

Responses recorded rostral to stimulation applied to the dorsal spinal cord are similar to those produced by peripheral stimulation (Figure 7). Additionally, a short latency response can be recorded from the anterior regions of the sensorimotor cortex secondary to spinal cord stimulation (Figure 8). This response is similar to that observed at the spinal cord with stimulation of the sensorimotor cortex and probably reflects antidromic conduction in the pyramidal tract.[57]

III. RESPONSE VARIATION WITH STIMULATION AND PHYSIOLOGIC PARAMETERS

A. Stimulus Intensity

The evoked potential with peripheral nerve stimulation is invariant with an increase in stimulus intensity above maximal motor threshold. However, for spinal stimulation, current spread into the motor fibers causes large motor responses. Consequently, the stimulus intensity must be monitored. In practice, the peak current intensity at a pulse duration of 0.2 msec for stimulation at SMC is typically 5 to 10 mA, while peripheral nerves require 2 to 6 mA and spinal cord requires 0.5 to 1.0 mA.

In contrast, the response in the intralaminar nuclei continues to increase in amplitude and complexity at stimulus intensities above motor threshold. Typically, pulse durations of 0.5 to 2 msec at intensities of 15 to 20 mA are used to obtain this response with peripheral nerve stimulation.

B. Stimulus Frequency

The responses from synaptic regions (dorsal root entry zone, dorsal column nuclei, VPL,

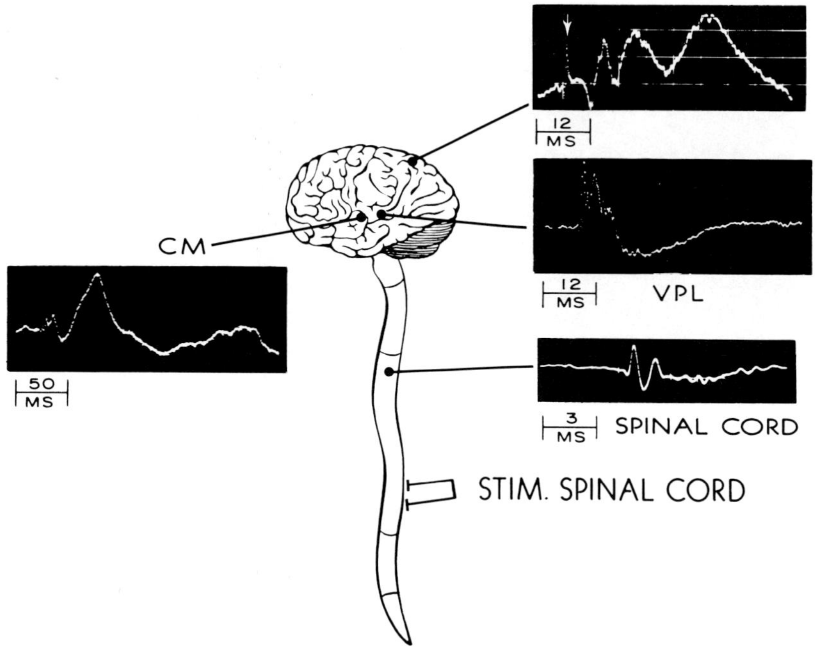

FIGURE 7. Responses obtained over dorsal spinal cord, sensorimotor cortex (top right), and thalamic nuclei VPL and CM with stimulation over the dorsal spinal cord of the monkey. Stimulus at arrow.

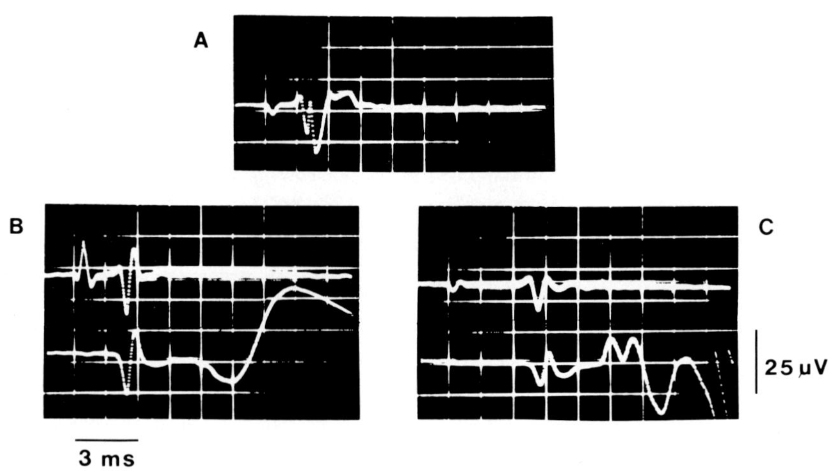

FIGURE 8. Comparison of responses recorded at spinal levels secondary to cortical stimulation, with those recorded at cortex secondary to spinal stimulation. Electrodes were placed over lower thoracic and lower cervical spinal cord and over sensorimotor cortex. (A) Response at cervical spinal cord with stimulation of thoracic spinal cord; (B) upper — response over cervical spinal cord with cortical stimulation; lower — response at cortex with stimulation of cervical spinal cord; (C) upper — response at thoracic spinal cord with stimulation of cortex; lower — response at cortex with thoracic spinal cord stimulation. Stimulus after first 10% of sweep.

and SMC) begin to diminish at frequencies of 10 to 20 Hz. The early portions of these responses are typically unaltered up to stimulus frequencies of 30 to 40 Hz. The dorsal

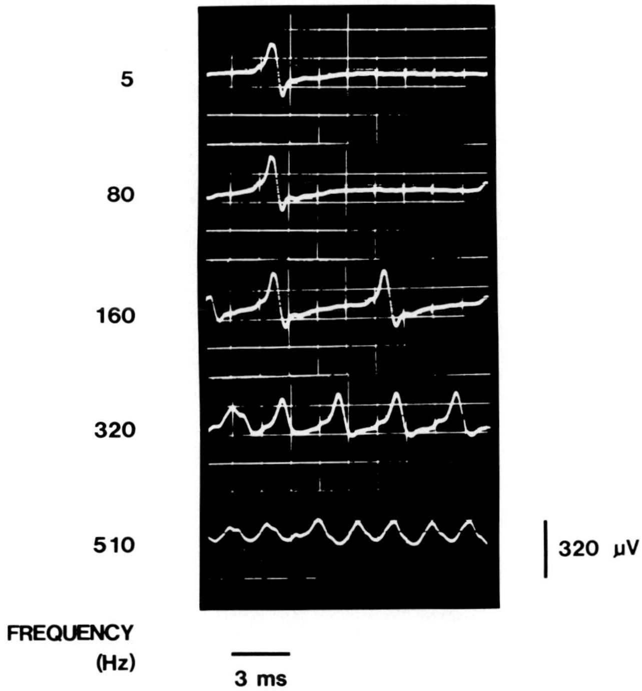

FIGURE 9. Effect of increasing stimulus frequency upon response recorded over the dorsal aspect of the cervical spinal cord with stimulation at thoracic spinal cord.

column nuclei response is not eliminated until approximately 60 to 70 Hz; however, VPL and SMC are essentially absent at stimulus frequencies of 50 to 60 Hz.

Potentials recorded on the spinal cord, with medial nerve, peroneal nerve, sensorimotor cortex, or spinal cord stimulation are unaltered up to approximately 200 Hz and often follow with diminished amplitude at stimulus frequencies of 600 to 700 Hz (Figure 9).

The response in thalamic intralaminar nuclei is generally reduced above 1 Hz and fails to follow at frequencies greater than 4 to 6 Hz (Figure 10). On stimulation of the intralaminar nuclei, a characteristic long-latency ''recruiting response'' is obtained at the cerebral cortex. Using averaging techniques, this response is not recorded until the stimulus frequency exceeds 5 to 6 Hz (Figure 6). It continues to increase with stimulation frequency up to 10 to 12 Hz.

C. Stimulus Duration

Typical stimulus durations for dorsal column system responses are 0.2 to 0.5 msec. With increasing stimulus pulse width above this level, no change is observed until the duration is equal to the refractory period for the response under study. With prolonged durations, an ''off response'' is observed. These responses are similar to those obtained with paired stimuli of opposite polarity with an interstimulus interval equal to the prolonged pulse width. The nonspecific responses continue to increase in amplitude and complexity with increasing pulse durations.

D. Anesthesia

Sedative doses of ketamine hydrochloride (5 mg/kg) administered intramuscularly have little affect on potentials recorded subdurally over the sensorimotor cortex (SMC) or over the dorsal column nuclei (DCN). Similarly, little alteration is observed for recordings obtained with bipolar depth electrodes in nucleus VPL of the thalamus. Responses obtained

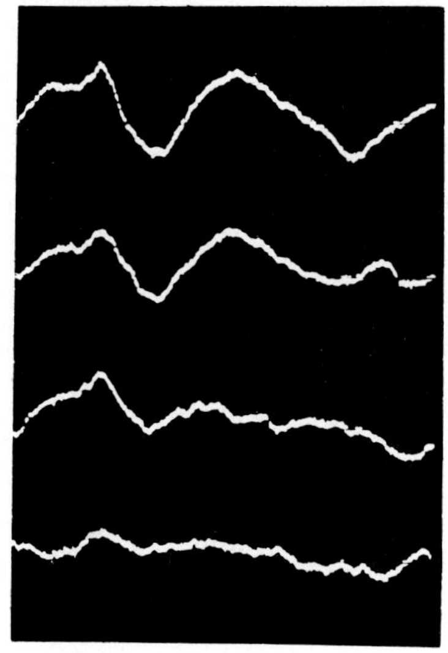

FIGURE 10. Effect of increasing stimulus frequency upon response recorded in thalamic intralaminar nucleus centre median secondary to peroneal nerve stimulation. Stimulus after first 10% of sweep.

with depth electrodes in the centre median or other intralaminar nuclei of the thalamus are essentially eliminated by ketamine or barbiturates. Anesthetic doses of i.v. ketamine enhance the early portions of the SMC response and attenuate later components. The effects last approximately 15 min from the time of injection.

As with other barbiturates, surgical doses of sodium thiamylal augment the earliest primary cortical inflection without latency changes and the longer latency cortical potentials are reduced (Figure 11). Volatile anesthetics, such as ether, halothane, and ethrane, eliminate the cortical evoked potential for up to 2 hr (Figure 12). For clinical procedures, nitrous oxide and oxygen are often used in conjunction with narcotics, because anesthesia can be maintained with little alteration in cortical evoked potentials.[75]

E. Ischemia

Afferent responses recorded rostral to the spinal cord are eliminated within 2 to 3 min of global ischemia. Similarly, responses of the efferent tracts, whether recorded at the spinal cord with cortical stimulation or recorded cortically with spinal stimulation, are obliterated rapidly. The long-duration part of the response at the dorsal root entry zone is also reduced promptly (Figure 13). The initial deflections of the evoked potential at conus medullaris and the thoracic cord responses to either peripheral nerve or spinal stimulation are unaltered up to 20 min following the onset of global ischemia. They then deteriorate over a 5- to 15-min interval and are completely eliminated within 25 to 35 min after cessation of flow.[53]

Most of the data concerning the relationship of the evoked potential and graded ischemia derive from the studies of Branston et al.[12-14] In these studies, utilizing a monopolar recording electrode in the skull, it was found that local tissue flows below 12 mℓ/100 g/min were correlated with abolition of the evoked potential while flows exceeding 16 mℓ/100 g/min left the response intact.

Contrary evidence is available from the studies of Bennett et al.[9] In these studies, it was

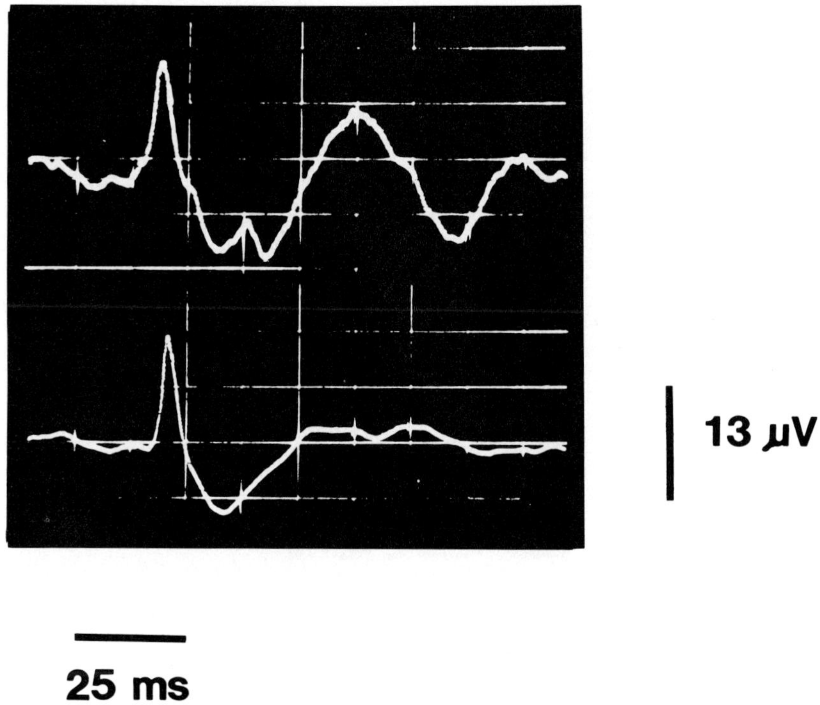

FIGURE 11. Effect of barbiturate (thiamylal sodium) on the response recorded at sensorimotor cortex with peroneal nerve stimulation: upper, control; lower, following i.v. administration of anesthetic. Stimulus at first 10% of sweep.

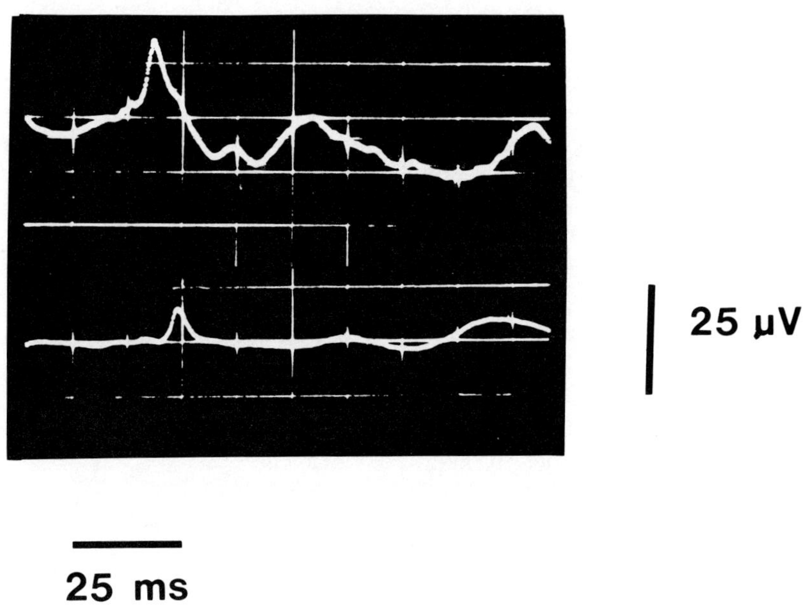

FIGURE 12. Effect of hologenated gas anesthetic on cortical evoked potential with peroneal nerve stimulation: upper, control; lower, during administration of anesthetic.

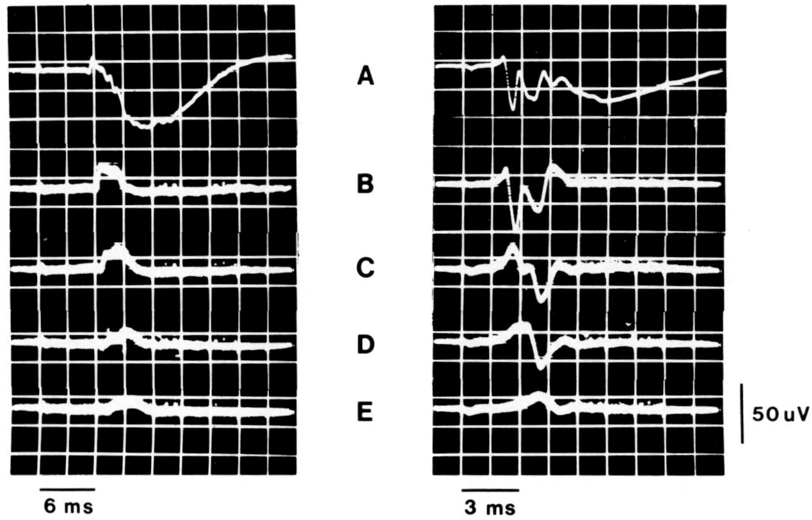

FIGURE 13. Responses at monkey conus medullaris (lumbar dorsal root entry zone) with stimulation of peroneal nerve (left) and over lower cervical spinal cord with stimulation of thoracic spinal cord (right). (A) Control; (B, C, D, and E) 5, 20, 23, and 25 min following cardiac arrest. Stimulus at first 10% of sweep.

found that a 20-mmHg pressure on the cortex produced a 9.8 to 14.1% reduction in rCBF and a 25 to 30% reduction in evoked potential amplitude.

Studies in our laboratory utilizing the combination hydrogen clearance isothermal flow probe[19,58] suggest that the cortical evoked potential is sensitive to cortical ischemia. The evoked potential began to be reduced when cerebral blood flow was reduced 10 to 20% and was obliterated at flows between 10 and 20 mℓ/100 g/min. Between these extremes, the response was reduced linearly with flow (Figure 14).

IV. SSEP IN NEURAL STIMULATION STUDIES

The somatosensory evoked potential has been used to experimentally evaluate neural stimulation,[7,8,11,32,33,48-50,52,54,59,66,68-71,77-81] spinal cord injury,[51,53,73-75] impact acceleration,[10,89] and epidural anesthetics.[20,21] The SSEP has proven to be a highly sensitive indicator of the functional integrity of the brain and spinal cord. Since neural systems and structures can be evaluated directly, the SSEP is an ideal tool for localizing the effects of experimental manipulations in animal models.

A. Spinal Stimulation

The somatosensory evoked potential (SSEP) has been applied to the experimental and clinical evaluation of peripheral nerve, spinal, cerebellar, and cerebral stimulation systems. Studies on spinal cord stimulation systems illustrate the use of the SSEP in the investigation of the effects of neural stimulation and the determination of preferred electrode position. SSEPs have been recorded in the monkey from sensorimotor cortex (SMC), thalamic nuclei ventralis posterior lateralis (VPL), and centre median (CM), midbrain reticular formation and the dorsal aspect of the spinal cord secondary to low frequency electrical stimulation of the spinal cord and peripheral nerve.

Three-in-line, five-in-line, and anterior to posterior electrode arrays have been used to apply high-frequency stimulation to the spinal cord.[48,49,68,82,83] The response at SMC, VPL, and spinal cord are obliterated by 100 Hz, 0.25-msec current pulses of approximately 0.2 to 0.5 mA applied to the 5-in-line electrodes placed over the dorsal columns between the

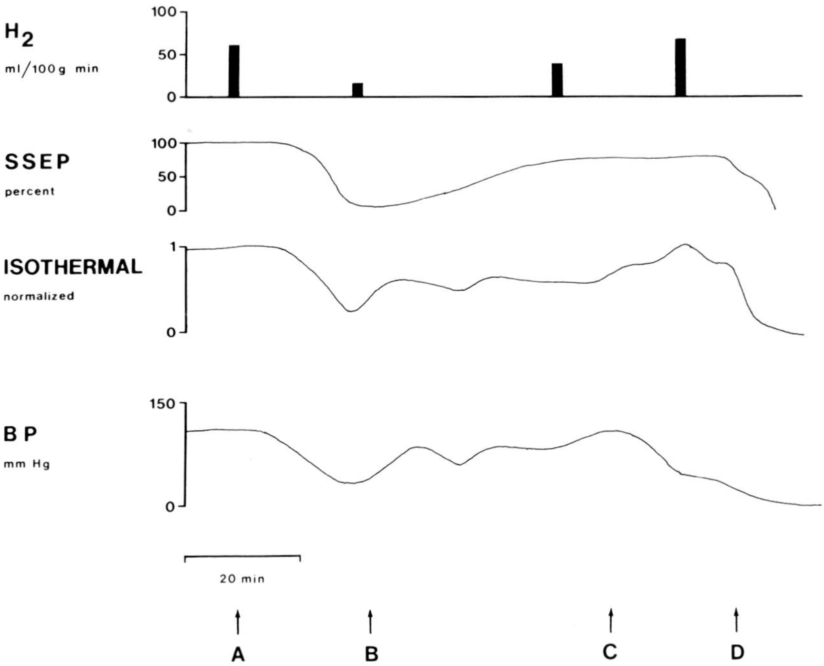

FIGURE 14. Comparison of normalized SSEP amplitude and local blood flow as obtained with the combination hydrogen clearance-isothermal flow probe. Mean arterial blood pressure is indicated in the lower trace. (A) Administration of nitroprusside; (B) stop nitroprusside; (C) restart nitroprusside; (D) cardiac arrest.

SSEP stimulation and recording sites (Figures 15 and 16). In contrast, when the currents are applied to electrodes on the anterior surface of the spinal cord, the responses in CM and the reticular formation are eliminated while those from the dorsal spinal cord, VPL, and SMC are intact. Currents passed from negative electrodes on the posterior surface to positive electrodes on the anterior surface of the spinal cord produce a reduction in the cortical SSEP at lower levels than the reverse polarity (Figure 17). If both the SSEP recording and stimulating electrodes are placed above or below the spinal cord stimulation, little or no effect is noted in the response.

When the currents are applied for brief periods, the SSEP recovers promptly. However, stimulation applied for periods greater than 3 to 5 min produces a prolonged conduction block.

Studies with various electrode configurations indicate that the current required to block the SSEP is proportional to the number of negative electrodes (approximately 0.11 to 0.15 mA per electrode). With current passed from positive electrodes on the spinal cord to a distal negative electrode, 0.3 to 0.4 mA is required. The average power density to block the SSEP was least with the five-electrode, alternate polarity ($+ - + - + -$) configuration (Table 1). In general, the power was least for those configurations with fewer negative than positive electrodes. Comparison of the monopolar configurations suggests that the block is produced at the negative electrode. Experiments on the effects of variation in pulse duration and frequency show that the current to suppress the evoked potential is inversely related to the duty cycle.

The effects of spinal cord stimulation upon pain are usually explained by the gate control theory proposed by Melzack and Wall.[55] Suppression of spinothalamic activity has been observed with isolated stimulation of the dorsal columns.[30] Although activation of excitatory and inhibitory systems is clearly an effect of the application of electrical currents on the

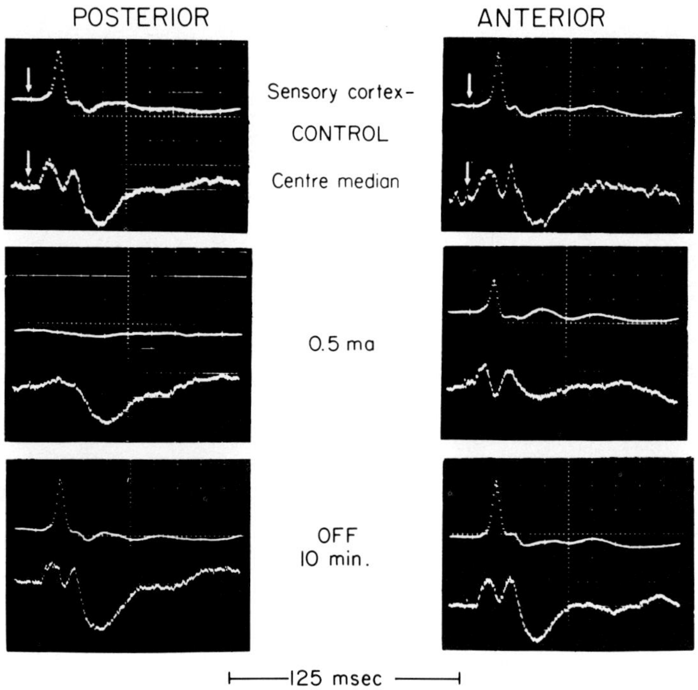

FIGURE 15. Responses recorded from monkey sensory cortex and nucleus centre median secondary to peripheral stimulation. Early deflection in CM response reflects activity in medial lemniscus. Recordings in left- and right-hand columns indicate effects of application of current at 100 Hz to posterior and anterior electrode sets, respectively. Arrow indicates shock artifact.[49]

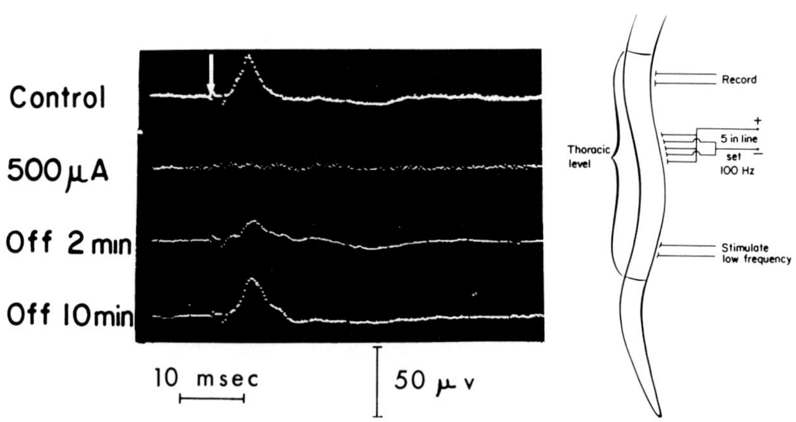

FIGURE 16. Recordings of cord-to-cord-evoked potentials before, during, and after current application to five-in-line electrodes on the dorsal surface of the cord. The stimulating and recording electrodes are separated by the five-in-line current electrodes. Barbiturate anesthesia, analysis time is 62 msec; stimulus at arrow. Response obliterated with 500 μA peak pulse current applied to 5-electrode set. Recovery at approximately 10 min after 100-Hz current is off.[48]

spinal cord, a direct block of neural transmission is also a possibility.[85] Since spinal cord stimulation produces a localized effect which is greatest at the negative electrode, greatest in the region of maximum current density, and is more dependent upon duty cycle than current level, a depolarizing mechanism must be considered.

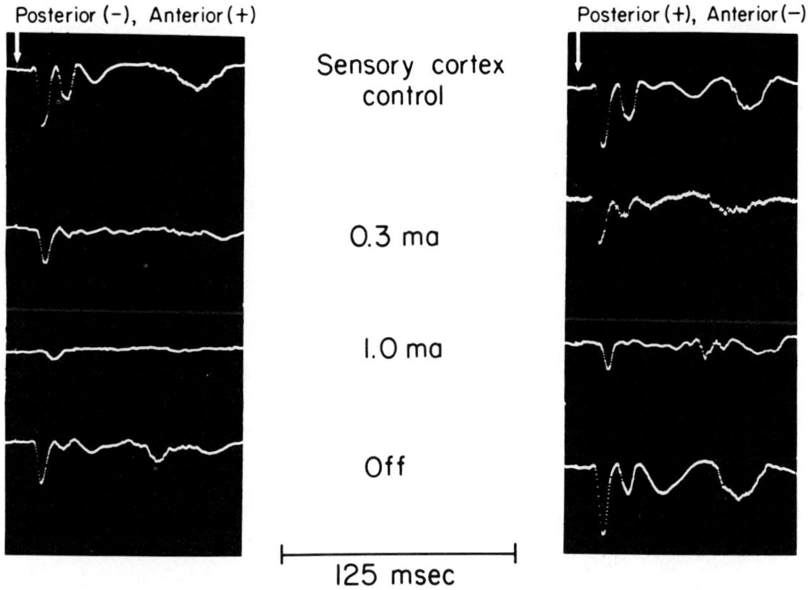

FIGURE 17. Tracings show effect of 100-Hz current polarity on monkey sensory cortex response to peripheral stimulation. Arrow indicates shock artifact. Responses on left were recorded during application of 100-Hz pulses of indicated amplitudes with all electrodes in posterior set connected to negative pole, while records on right were made when posterior set was positive. Bottom recording in each column was made just after 100-Hz currents were stopped.[49]

The results of these animal studies correlate with clinical experience.[54] Patients with dorsal column electrodes routinely experience paresthesias and reduced joint perception at and below currents required for pain relief. The SSEP was reduced at currents of approximately 0.5 to 1.0 mA. With anterior column stimulation, pain relief was obtained without reduction of the evoked potential or dorsal column sensory effects. When currents were passed from anterior to posterior, pain relief without sensory side effects was obtained at lower levels (approximately 0.25 mA). Additionally, no increase in applied current is required over time, in contrast to the single electrode arrays.

Neural stimulation often affords unique opportunities to study the function of the human CNS.[34,48,62,79,80] Figure 18 was obtained during implantation of electrodes on the anterior aspect of the spinal cord. No cortical response was observed until the applied current magnitude was sufficient to stimulate the dorsal columns.[45]

B. Cerebellar Stimulation

The relationship between the clinical effectiveness of cerebellar stimulation, and its effect on the evoked potential has not been established. However, reduction of the SSEP has usually been observed coincident with improved function.[50,52,86] Consequently, studies were conducted in the chimpanzee with large arrays of individually accessible electrodes (Figure 19) to determine the electrode configuration and stimulus parameter which had the greatest effect on the SSEP.

Substantial reduction of the SSEP was accomplished with the least current density when all of the cerebellar electrodes were negative with respect to a distal electrode (Figure 20). This effect is probably due to the large surface area affected with this configuration. The difference between the totally monopolar configuration and that in which the current is passed across the cerebellum is probably due to the directon of the current flow.[83]

Table 1
POWER DENSITY AT ELECTRODE FOR VARIOUS POLARITY CONFIGURATIONS OF FIVE-ELECTRODE SYSTEM

Configuration	Maximum density of average power (mW/cm^2)
+ − + − +	0.0036
− + − + −	0.0086
+ 0 − 0 +	0.056
− 0 + 0 −	0.013
0 + − + 0	0.0056
0 − + − 0	0.013
+ − + 0 0	0.0059
0 0 0 − 0	0.0042
0 − − − 0	0.0055
0 + + + 0	0.0402
− − − − −	0.0058
+ + 0 − −	0.0042

Note: Sufficient current was applied to suppress evoked potentials. Power was calculated by numerical approximation of the formula below and normalized with respect to electrode area.

$$P_{Ave} = \frac{1}{T}\int_0^T v(t)i(t)dt$$

where v(t) is applied voltage (V), i(t) is applied current (mA), T is period of v(t) and i(t), and P_{Ave} is power averaged over the period T (mW).

Reduction of subcortical evoked potentials has also been observed with cerebellar stimulation.[11,86] This, together with the fact that reductions in spasticity are observed without an intact cortex,[39] suggest that cerebellar stimulation acts by way of a thalamic mechanism.

Current application between the superior to posterior electrode sets of 1 cerebellar hemisphere showed that the average power and charge density per negative electrode for a 50% reduction of amplitude of the SSEP was least near 10 and 100 Hz. The power increased with pulse width from 0.1 to 1 msec (Figure 21). Tests over a 4-week period were repeatable within 20 to 30%. For pulse durations shown, a maximum reduction of 60% in the SSEP amplitude was observed at 10 Hz. However, the SSEP could be obliterated at frequencies above 50 Hz.

Similar results were obtained when all 27 electrodes of the same hemisphere were connected to the negative pole of the generator. However, the average power levels and charge density were approximately 50% less for a 50% reduction in SSEP amplitude. All values are 1.7 times greater for the positive electrodes of the posterior cerebellar surface. For these studies, the charge density per pulse and peak current density per pulse per negative electrode decreased approximately linearly from 10 to 100 Hz. A slower change was observed above 100 Hz.

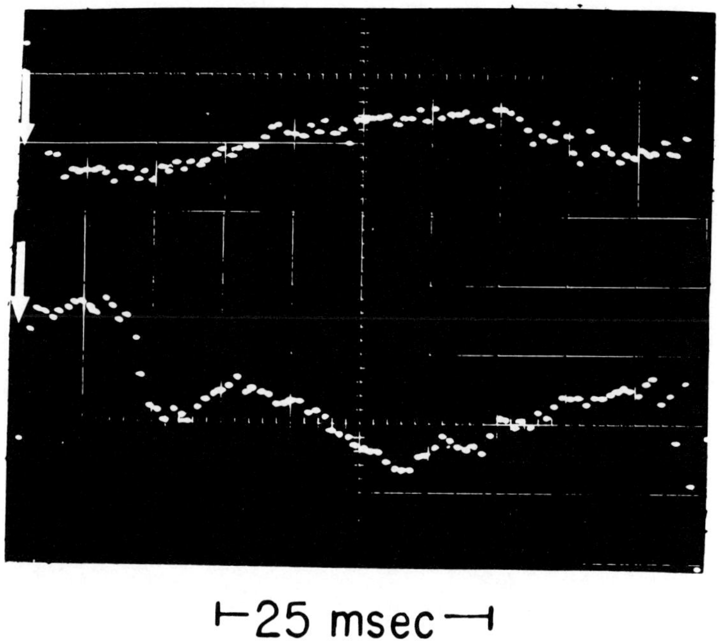

FIGURE 18. Recordings from scalp electrodes in patient, with stimulation through anterior electrode set at 10 Hz. For top tracing, level of current used was same as that required to produce analgesia and pain relief when applied at 100 Hz. Bottom recording was made after level of current applied at 10 Hz had been substantially increased. Arrows indicate shock artifact.[49]

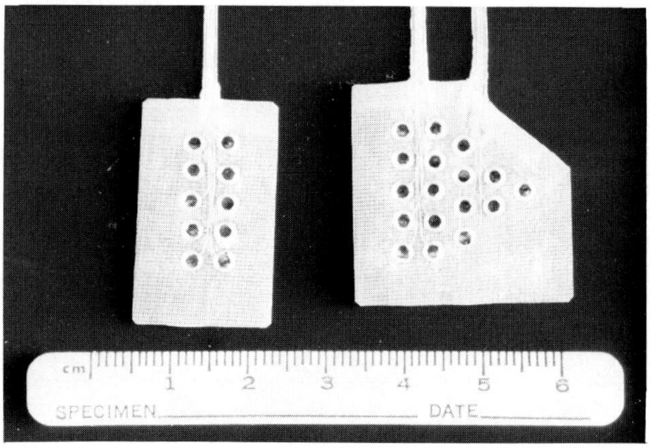

FIGURE 19. All electrodes are 2-mm-diameter platinum discs, each attached to a separate flexible stainless steel. Teflon®-coated cable is connected to an Amphenol plug imbedded in the calvarium. The left electrode array is implanted subdurally on the posterior surface of the chimpanzee cerebellum; the right array is implanted subdurally over the superior surface of the cerebellum.[70]

ACKNOWLEDGMENT

This research was supported in part by the Office of Naval Research Contract No. N0014-77-C-0749 and the Veterans Administration Medical Center Research Grant #6000-02P.

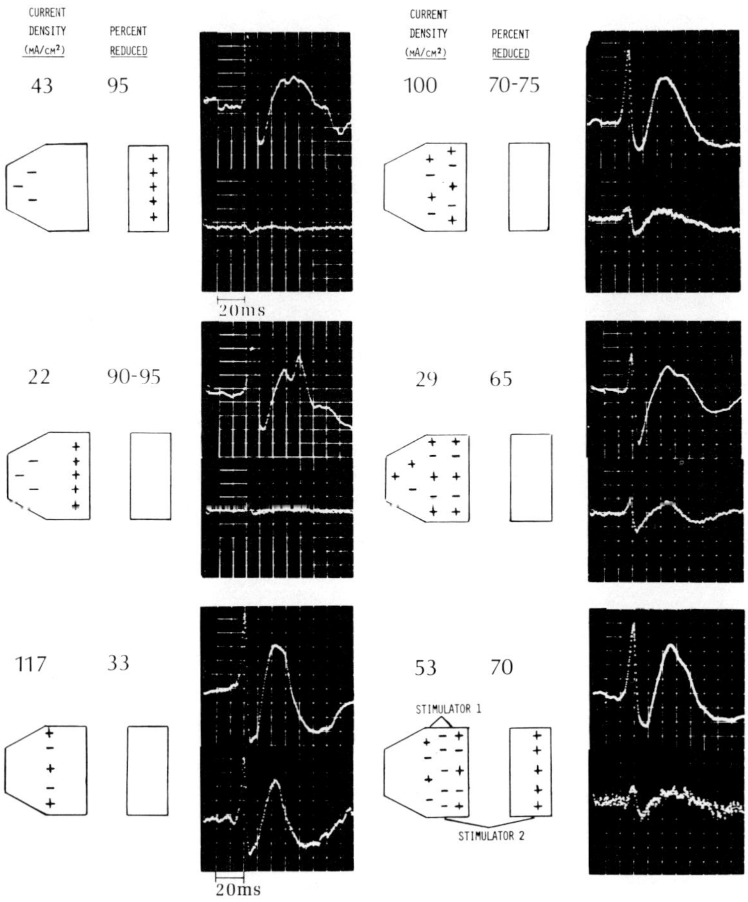

FIGURE 20. Typical evoked potential recordings at the contralateral sensorimotor cortex secondary to transcutaneous electrical nerve stimulation at the leg of the chimpanzee. Top record is control, and lower is with current applied. The polarity and electrodes used in the study are indicated to the left of the evoked potentials; the current density (mA/cm^2) is given in the column on the left in each instance; the percentage reduction in the early component of the evoked potential is given on the right.[71]

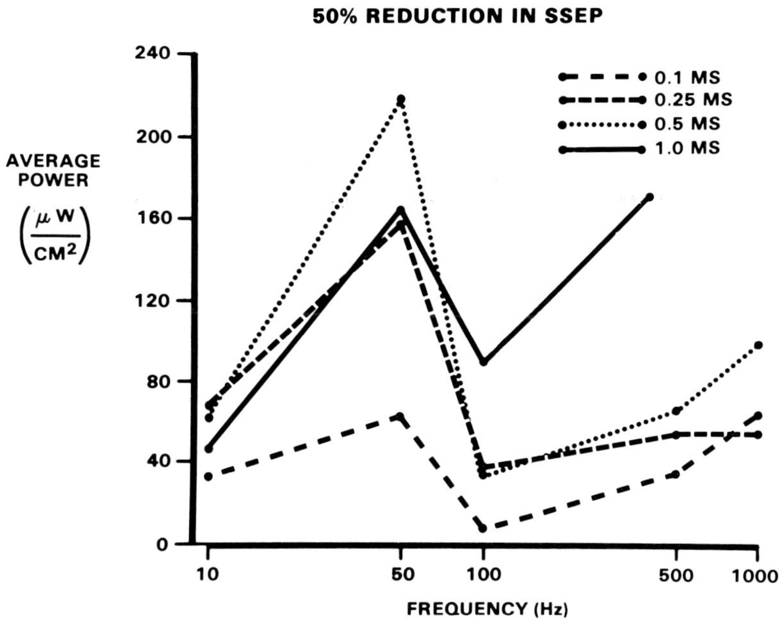

A

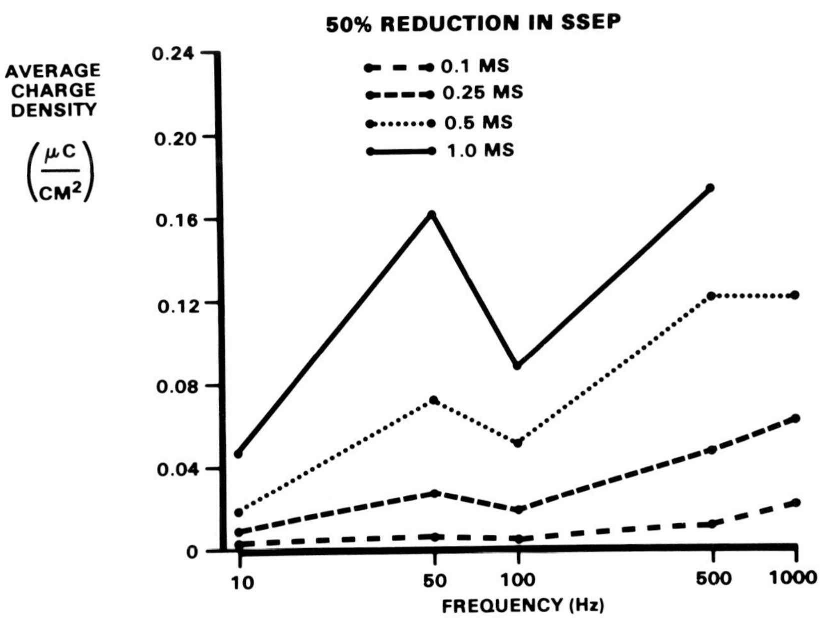

B

FIGURE 21. (A) Average power density in W/cm^2 per negative electrode vs. frequency for a 50% reduction in amplitude of the somatosensory evoked potential. The somatosensory evoked potential was measured at sensorimotor cortex with application of 4-Hz, 0.2-msec duration supramaximal stimulus to the peroneal nerve of the leg contralateral to the recording site. Separate plots are made for 0.1-, 0.25-, 0.5-, and 1.0-msec-duration capacitively coupled pulses. (B) Average charge density in per square centimeter (C/cm^2) per negative electrode vs. frequency for a 50% reduction in amplitude of the somatosensory evoked potential. (C) Charge density per pulse in C/cm^2 per negative electrode vs. frequency for a 50% reduction in amplitude of the somatosensory evoked potential. These are the corresponding curves from data used to calculate (A). (D) Peak current density per pulse in C/cm^2 per negative electrode vs. frequency for a 50% reduction in amplitude of the somatosensory evoked potential. These are the corresponding curves from data used to calculate (B).[71]

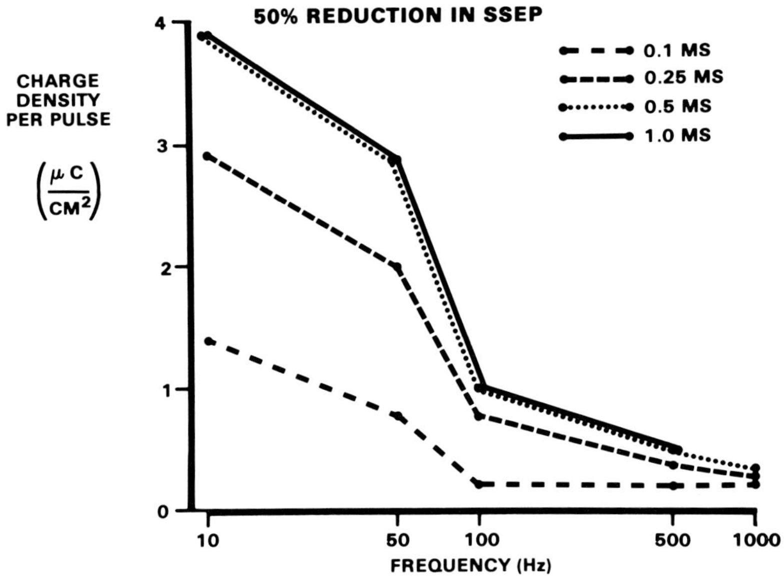

FIGURE 21C.

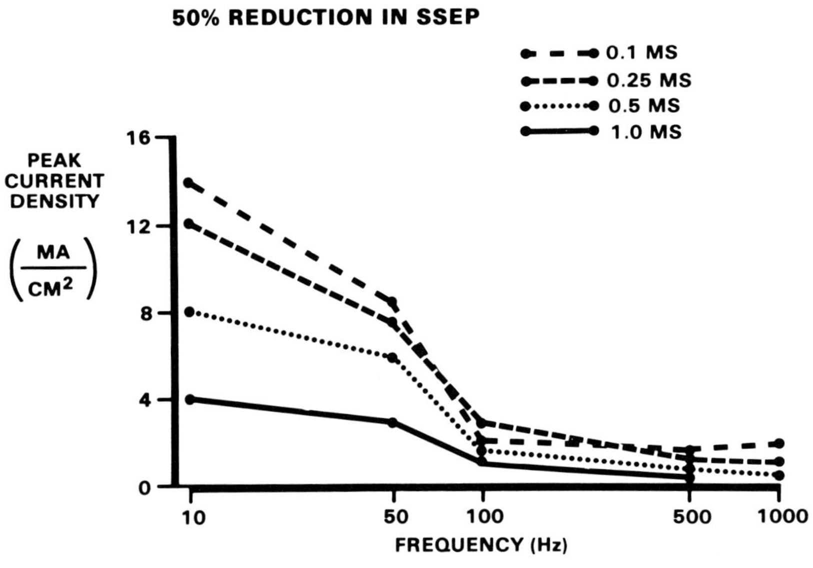

FIGURE 21D.

REFERENCES

1. **Ackmann, J. J.,** A computer system for neurosurgical patient monitoring, *Comput. Programs Biomed.*, 10, 81, 1979.
2. **Ackmann, J. J., Larson, S. J., Sances, A., Jr., and Barr, R. E.,** Non-invasive monitoring techniques in neurosurgical intensive care, *J. Clin. Eng.*, 4(4), 329, 1979.
3. **Ackmann, J. J., Larson, S. J., Sances, A. Jr., Hemmy, D. C., Cusick, J. F., and Evans, S. M.,** An automated system of monitoring patients with head injury, in *Electrotherapeutic Sleep and Electroanesthesia*, Vol. 5, Wageneder, F. M. and Germann, R. H., Eds., University of Graz, Graz, Austria, 1978, 19.
4. **Anziska, B. J. and Cracco, R. Q.,** Short latency somatosensory evoked potentials in brain dead patients, *Arch. Neurol.*, 37, 222, 1980.
5. **Austin, G. M. and McCouch, G. P.,** Presynaptic component of intermediary cord potential, *J. Neurophysiol.*, 18, 441, 1955.
6. **Ball, G. J., Sanders, M. G., and Schnabl, J.,** Determination of peripheral sensory nerve conduction velocities in man from stimulus response delays of the cortical evoked potentials, *Electroencephalogr. Clin. Neurophysiol.*, 30, 409, 1971.
7. **Bantli, H., Bloedel, J. R., Long, D. M., and Thienprasit, P.,** Distribution of activity in spinal pathways evoked by experimental dorsal column stimulation, *J. Neurosurg.*, 42, 290, 1975.
8. **Bantli, H., Bloedel, J. R., and Thienprasit, P.,** Supraspinal interactions resulting from experimental dorsal column stimulation, *J. Neurosurg.*, 42, 296, 1975.
9. **Bennett, M. H., Albin, M. S., Bunegin, L., et al.,** Evoked potential changes during brain retraction in dogs, *Stroke*, 8(4), 487, 1977.
10. **Berger, M. and Weiss, M.,** Effects of impact acceleration on somatosensory evoked potentials, in *Impact Injury of the Head and Spine*, Ewing, C. L., Thomas, D. J., Sances, A., Jr., and Larson, S. J., Eds., Charles C Thomas, Springfield, Ill., 1983, 324.
11. **Boone, S. C., Nashold, B. S., and Wilson, W. P.,** The effects of cerebellar stimulation on the averaged sensory evoked responses in the cat, in *Cerebellum, Epilepsy and Behavior*, Cooper, I. S., Riklan, M., and Snider, R. S., Eds., Plenum Press, New York, 1973, 257.
12. **Branston, N. M., Symon, L., Crockard, H. A., et al.,** Relationship between the cortical evoked potential and local cortical blood flow following acute middle cerebral artery occlusion in the baboon, *Exp. Neurol.*, 45, 195, 1974.
13. **Branston, N. M., Symon, L., and Crockard, H. A.,** Recovery of the cortical evoked response following middle cerebral artery occlusion in baboons — relation to local flow and pO2, *Stroke*, 7, 151, 1976.
14. **Branston, N. M., Strong, A., and Symon, L.,** Extracellular potassium activity, evoked potential and tissue blood flow, *J. Neurol. Sci.*, 32, 305, 1977.
15. **Cummins, K. L., Perkel, D. H., and Dorfman, L. J.,** Nerve fiber conduction-velocity distributions. I. Estimation based on the single fiber and compound action potentials, *Electroencephalogr. Clin. Neurophysiol.*, 46, 634, 1979.
16. **Cusick, J. F., Myklebust, J., Larson, S. J., and Sances, A. J.,** Spinal evoked potentials in the primate: neural substrate, *J. Neurosurg.*, 49(4), 551, 1978.
17. **Cusick, J. F., Myklebust, J. B., Larson, S. J., and Sances, A., Jr.,** Spinal cord evaluation by cortical evoked responses, *Arch. Neurol.*, 36(3), 140, 1979.
18. **Cusick, J. F., Ackmann, J. J., and Larson, S. J.,** Mechanical and physiological effects of dentatomy, *J. Neurosurg.*, 46, 767, 1977.
19. **Cusick, J. F. and Myklebust, J.,** Continuous quantitative local cerebral blood flow measurement. Calibration of thermal conductivity measurements by the hydrogen clearance method, *Stroke*, 11(6), 661, 1980.
20. **Cusick, J. F., Myklebust, J. B., and Abram, S. E.,** Differential neural effects of epidural anesthetics, *Anesthesiology*, 53(4), 299, 1980.
21. **Cusick, J. F., Myklebust, J. B., Abram, S. E., and Davidson, A.,** Altered neural conduction with epidural bupivacaine, *Anesthesiology*, 57, 31, 1982.
22. **de la Torre, J. C., Trimble, J. L., Beard, R. T., Hanlon, K., and Surgeon, J.,** Somatosensory evoked potentials for the prognosis of coma in humans, *Exp. Neurol.*, 60, 304, 1978.
23. **Delbeke, J., McComas, A. J., and Kopec, S. J.,** Analysis of evoked lumbosacral potentials in man, *J. Neurol. Neurosurg. Psychiatry*, 41, 293, 1978.
24. **Dempsey, E. W. and Morison, R. S.,** The production of rhythmically recurrent cortical potentials after localized thalamic stimulation, *Am. J. Physiol.*, 135, 293, 1942.
25. **Dempsey, E. W. and Morison, R. S.,** The interaction of certain spontaneous and induced cortical potentials, *Am. J. Physiol.*, 135, 301, 1942.
26. **Desmedt, J. E., Mantil, J., Borenstein, S., Debecker, J., Lambert, C., Franken, L., and Danis, A.,** Evaluation of sensory nerve conduction from averaged cerebral evoked potentials in neuropathies, *Electromyogr. Clin. Neurophysiol.*, 6, 263, 1966.

27. **Eccles, J. C., Kostyuk, P. G., and Schmidt, R. F.,** Central pathways responsible for depolarization of primary afferent fibers, *J. Physiol.*, 161, 237, 1962.
28. **Elliot, L., Halliday, R., and Callaway, E.,** Brain event related potentials: contributions to research in learning disabilities, in *Progress in Learning Disabilities*, Myklebust, H. R., Ed., Grune & Stratton, New York, 1978, 121.
29. **Engler, G. L., Spielholz, N. I., Bernhard, W. N., Danziger, F., Merkin, H., and Wolff, T.,** Somatosensory evoked potentials during harrington instrumentation for scoliosis, *J. Bone Jt. Surg.*, 60, 528, 1978.
30. **Foreman, R. D., Beall, J. E., Applebaum, A. E., Coulter, J. D., and Willis, W. D.,** Effects of dorsal column stimulation on primate spinothalamic tract neurons, *J. Neurophysiol.*, 39, 534, 1976.
31. **Giblin, D. R.,** Somatosensory evoked potentials in healthy subjects and in patients with lesions of the nervous system, *Ann. N.Y. Acad. Sci.*, 112, 93, 1964.
32. **Gildenberg, P. L. and Murthy, K. S. K.,** Modification of thalamic evoked activity by dorsal column stimulation, *Acta Neurochir. Suppl.*, 24, 159, 1977.
33. **Gildenberg, P. L. and Murthy, K. S. K.,** Influence of dorsal column stimulation upon human thalamic somatosensory-evoked potentials, *Appl. Neurophysiol.*, 43, 8, 1980.
34. **Gottlieb, G. L. and Myklebust, B. M.,** Some effects of cerebellar stimulation of cerebral palsy patients: changes in spinal reflexes and ankle joint compliance, in *Neural Stimulation*, Vol. 2, Myklebust, J. B., Sances, A., Jr., Larson, S. J., and Cusick, J. F., Eds., CRC Press, Boca Raton, Fla., 1985, chap. 4.
35. **Greenberg, R. P., Becker, D. P., Miller, J. D., and Mayer, D. J.,** Evaluation of brain function in severe human head trauma with multimodality evoked potentials. II, *J. Neurosurg.*, 47, 163, 1977.
36. **Greenberg, R. P., Mayer, D. J., Becker, D. P., and Miller, J. D.,** Evaluation of brain function in severe human head trauma with multimodality evoked potentials, I, *J. Neurosurg.*, 47, 150, 1977.
37. **Halliday, A. M. and Wakefield, G. S.,** Cerebral evoked potentials in patients with dissociated sensory loss, *J. Neurol. Neurosurg. Psychiatry*, 26, 211, 1963.
38. **Hattori, S., Saiki, K., and Kawai, S.,** Diagnosis of the level and severity of cord lesion in cervical spondylotic myelopathy, *Spine*, 4(6), 478, 1979.
39. **Hemmy, D. C., Larson, S. J., Sances, A., Jr., and Millar, E. A.,** The effect of cerebellar stimulation on focal seizure activity and spasticity in monkeys, *J. Neurosurg.*, 46, 648, 1977.
40. **Ishijima, B., Yoshimasu, N., and Fukushima, T.,** Nociceptive neurons in the human thalamus, *Confin. Neurol.*, 37, 99, 1975.
41. **Kritchevsky, M. and Wiederholt, W. C.,** Short-latency somatosensory evoked potentials, *Arch. Neurol.*, 35, 706, 1978.
42. **Larson, S. J., Sances, A., Jr., and Baker, J. B.,** Evoked cortical potentials in patients with stroke, *Circulation*, 23 (Suppl. II), 15, 1966.
43. **Larson, S. J. and Sances, A., Jr.,** Evoked potentials in man: neurosurgical applications, *Am. J. Surg.*, 111, 857, 1966.
44. **Larson, S. J., Sances, A., Jr., and Christenson, P. C.,** Evoked somatosensory potentials in man, *Arch. Neurol.*, 15, 88, 1966.
45. **Larson, S. J. and Sances, A., Jr.,** Averaged evoked potentials in stereotaxic surgery, *J. Neurosurg.*, 28(3), 227, 1968.
46. **Larson, S. J. and Sances, A., Jr.,** The specific somatosensory system and dyskinesia, *Arch. Neurol.*, 18, 543, 1968.
47. **Larson, S. J., Sances, A., Jr., Ackmann, J. J., and Riegel, D. H.,** Non-invasive evaluation of head trauma patients, *Surgery*, 74, 34, 1973.
48. **Larson, S. J., Sances, A., Jr., Riegel, D. H., et al.,** Neurophysiological effects of dorsal column stimulation in man and monkey, *J. Neurosurg.*, 41(2), 217, 1974.
49. **Larson, S. J., Sances, A., Jr., Cusick, J. F., et al.,** A comparison between anterior and posterior spinal implant systems, *Surg. Neurol.*, 4, 180, 1975.
50. **Larson, S. J., Sances, A., Jr., Cusick, J. F., Myklebust, J., Millar, E. A., Boehmer, R., Hemmy, D. C., and Ackmann, J. J.,** Cerebellar implant studies, *IEEE Trans. Biomed. Eng.*, 23, 319, 1976.
51. **Larson, S. J., Holst, R. A., Hemmy, D. C., and Sances, A., Jr.,** The lateral extracavitary approach to traumatic lesions of the thoracic and lumbar spine, *J. Neurosurg.*, 45, 628, 1976.
52. **Larson, S. J., Sances, A., Jr., Hemmy, D. C., Millar, E. A., and Walsh, P. R.,** Physiological and histological effects of cerebellar stimulation, *Appl. Neurophysiol.*, 40(2-4), 160, 1977/78.
53. **Larson, S. J., Walsh, P. R., Sances, A., Jr., Cusick, J. F., Hemmy, D. C., and Mahler, H.,** Evoked potentials in experimental myelopathy, *Spine*, 5(4), 299, 1980.
54. **Larson, S. J., Sances, A., Jr., and Maiman, D. J.,** Spinal cord stimulation for pain, in *Neural Stimulation*, Vol. 1, Myklebust, J. B., Sances, A., Jr., Larson, S. J., and Cusick, J. F., Eds., CRC Press, Boca Raton, Fla., 1985, chap. 8.
55. **Melzack, R. and Wall, P. D.,** Pain mechanisms: a new theory, *Science*, 150, 972, 1965.

56. **Morison, R. S. and Dempsey, E. W.,** A study of thalamo-cortical relations, *Am. J. Physiol.*, 135, 281, 1942.
57. **Myklebust, J., Sances, A., Jr., Cusick, J., and Larson, S.,** Evoked potentials with spinal stimulation, *Proc. 11th Ann. Soc. Neuroscience*, 7, 284, 1981.
58. **Myklebust, J. B., Cusick, J. F., Sances, A., Jr., Salles-Cunha, S. X., and Macur, R. A.,** A combination isothermal-hydrogen clearance system for the measurement of local tissue flow, *IEEE Trans. Biomed. Eng.*, 28(3), 265, 1981.
59. **Myklebust, J., Sances, A., Jr., Swiontek, T., et al.,** Spinal cord stimulation studies, *Proc. 6th Int. Symp. on Electrostimulation*, Albena, Bulgaria, September 24, 1981.
60. **Namerow, N. S.,** Somatosensory evoked responses in multiple sclerosis patients with varying sensory loss, *Neurology*, 18, 1197, 1968.
61. **Narikashvili, S. P.,** Influence of unspecific impulses on the responses of sensory cortex, in *Progress in Brain Research*, Vol. 1, American Elsevier, New York, 1963, 155.
62. **Nashold, B., Somjen, G., and Friedman, H.,** Paresthesias and EEG potentials evoked by stimulation of the dorsal funiculi in man, *Exp. Neurol.*, 36, 273, 1972.
63. **Nathan, P. W.,** Pain, *Br. Med. Bull.*, 33(2), 149, 1977.
64. **Noel, P. and Desmedt, J. E.,** Somatosensory cerebral evoked potentials after vascular lesions of the brainstem and diencephalon, *Brain*, 98, 113, 1975.
65. **Purpura, D. P., Scarff, T., and McMurtry, J. G.,** Intracellular study of internuclear inhibition in ventrolateral thalamic neurons, *J. Neurophysiol.*, 28, 487, 1965.
66. **Salar, G., Iob, I., and Mingrino, S.,** Cortical evoked responses and transcutaneous electrotherapy, *Neurology*, 30, 663, 1980.
67. **Sances, A., Jr. and Larson, S. J.,** Evoked potential recording: an adjunct to human stereotactic surgery, *IEEE Trans. Biomed. Eng.*, 14, 162, 1967.
68. **Sances, A., Jr., Swiontek, T. J., Larson, S. J., et al.,** Innovations in neurologic implant systems, *Med. Instrum.*, 9(5), 213, 1975.
69. **Sances, A., Jr. and Larson, S. J.,** *Electroanesthesia: Biomedical and Biophysical Studies*, Academic Press, New York, 1975.
70. **Sances, A., Jr., Larson, S. J., Myklebust, J., Swiontek, T., Millar, E., Cusick, J. F., Hemmy, D. C., Jodat, R., and Ackmann, J. J.,** Studies of electrode configuration (Symposium: Neuroaugmentive Devices, Session on Chronic Cerebellar Stimulation), *Neurosurgery*, 1(2), 207, 1977.
71. **Sances, A., Jr., Larson, S. J., Myklebust, J., Swiontek, T., Millar, E. A., Cusick, J. F., Hemmy, D. C., Jodat, R., and Ackmann, J. J.,** Evaluation of electrode configurations in cerebellar implants, *Appl. Neurophysiol.*, 40(2-4), 141, 1977/78.
72. **Sances, A., Jr., Myklebust, J., Larson, S. J., Cusick, J. F., and Walsh, P. R.,** Theoretical investigations and clinical applications of the evoked potential, *IEEE 1979 Frontiers of Engineering in Health Care*, IEEE, Piscataway, N.J., 1979, 228.
73. **Sances, A., Jr., Weber, R., Myklebust, J., Cusick, J. F., Larson, S., Walsh, P., Christoffel, T., Houterman, C., Ewing, C., Thomas, D., and Saltzberg, B.,** The evoked potential: an experimental method for biomechanical analysis of brain and spinal injury, *SAE Trans.*, 89, 3815, 1980.
74. **Sances, A., Jr., Myklebust, J., Cusick, J., Weber, R., Houterman, C., Larson, S., Walsh, P., Chilbert, M., et al.,** Experimental studies of brain and neck injury, in *Proc. 25th Stapp Car Crash Conf.*, Society of Automotive Engineers, Warrendale, Pa., 1981, 149.
75. **Sances, A., Jr., Myklebust, J., Larson, S. J., Cusick, J. F., and Weber, R.,** The evoked potential — a biomechanical tool, in *Impact Injury of the Head and Spine*, Ewing, C. L., Thomas, D. J., Sances, A., Jr., and Larson, S. J., Eds., Charles C Thomas, Springfield, Ill., in press.
76. **Sarnowski, R. J., Cracco, R. Q., Vogel, H. B., and Mount, F.,** Spinal evoked response in the cat, *J. Neurosurg.*, 43, 329, 1975.
77. **Sedgwick, E. M., Illis, L. S., Tallis, R. C., Thornton, A. R. D., Abraham, P., El-Negamy, E., Docherty, T. B., Soar, J. S., Spencer, S. C., and Taylor, F. M.,** Evoked potentials and contingent negative variation during treatment of multiple sclerosis with spinal cord stimulation, *J. Neurol. Neurosurg. Psychiatry*, 43, 15, 1980.
78. **Sedgwick, E. M., Thornton, A. R. D., El-Negamy, E., Jobbling, D. T., Tallis, R. C., and Illis, L. S.,** Electrophysiological responses and stimulus parameters associated with spinal cord stimulation, *Int. Rehab. Med.*, 2(2), 71, 1980.
79. **Sherwood, A. M.,** Characteristics of somatosensory evoked potentials recorded over the spinal cord and brain of man, *IEEE Trans. Biomed. Eng.*, 28(7), 481, 1981.
80. **Sherwood, A. M.,** Electrical stimulation of the spinal cord in movement disorders, in *Neural Stimulation*, Vol. 1, Myklebust, J. B., Sances, A., Jr., Larson, S. J., and Cusick, J. F., Eds., CRC Press, Boca Raton, Fla., 1984, chap. 7.

81. **Shimoji, K., Matsuki, M., Shimizu, H., Iwane, T., Takahashi, R., Maruyama, M., and Masuko, K.,** Low-frequency, weak extradural stimulation in the management of intractable pain, *Br. J. Anaesth.*, 49(11), 1081, 1977.
82. **Swiontek, T. J., Sances, A., Jr., Larson, S. J., et al.,** Spinal cord implant studies, *IEEE Trans. Biomed. Eng.*, 23(4), 307, 1976.
83. **Swiontek, T. and Sances, A., Jr.,** Experimental current density measurement, in *Neural Stimulation*, Vol. 1, Myklebust, J. B., Sances, A., Jr., Larson, S. J., and Cusick, J. F., Eds., CRC Press, Boca Raton, Fla., 1985, chap. 4.
84. **Toth, S., Vajda, J., and Solyom, A.,** The function of the human cerebellum studied by evoked potentials and motor reactions, *Appl. Neurophysiol.*, 43, 48, 1980.
85. **Torebjork, H. E. and Hallin, R. G.,** Responses in human A and C fibres to repeated electrical intradermal stimulation, *J. Neurol. Neurosurg. Psychiatry*, 37, 653, 1974.
86. **Upton, A. R. and Cooper, I. S.,** Some neurophysiological effects of cerebellar stimulation in man, *Can. J. Neurol. Sci.*, 3, 237, 1976.
87. **Urabe, M. and Tsubokawa, T.,** Stereotaxic thalamotomy for the relief of intractable pain — CEM thalamotomy, *Tohoku J. Exp. Med.*, 85, 286, 1965.
88. **Wall, P. D.,** Excitability change in afferent fibre terminations and their relation to slow potentials, *J. Physiol.*, 142, 1, 1958.
89. **Walsh, P. R. and Jessop, M. E.,** The evoked potential in sled impact acceleration: methodologic and neurosurgical considerations, in *Impact Injury of the Head and Spine*, Ewing, C. L., Thomas, D. J., Sances, A., Jr., and Larson, S. J., Eds., Charles C Thomas, Springfield, Ill., in press.

Chapter 6

SELECTIVE ELECTRICAL ACTIVATION OF AFFERENTS: A TOOL TO INVESTIGATE NEURAL INTEGRATION

Edward J. Zuperku and Francis A. Hopp

TABLE OF CONTENTS

I. INTRODUCTION

Reflex regulation provides a mechanism whereby a biological organism is capable of maintaining vital stability in its internal environment in the face of sudden changes in external environment or changes brought about by volitional action. The compromise of some aspect of this regulation, which may result from disease, injury, or anesthesia, may render the system incapable of responding appropriately to a stressful situation. Therefore a better understanding of the importance and limitations of reflex mechanisms in regulation is of interest.

A representative example of regulation is that which involves pulmonary ventilation. Pulmonary ventilation is controlled so as to maintain the values of arterial PO_2, PCO_2, and pH within a narrow range. The breathing pattern which produces the required ventilation is reflexly controlled on a breath-to-breath basis. Various types of mechano- and chemoreceptors supply afferent input to the respiratory center, which when integrated, have a profound effect on the inspiratory duration, T_I, expiratory duration, T_E, and the time-course of respiratory motoneuron output. It is thought that the resultant breathing pattern is based on a performance criterion which is associated with the effectiveness of oxygen transport, the removal of metabolically produced CO_2, and the minimization of energy consumption.

While many recent studies have contributed to a better understanding of the reflex control of breathing, details of the mechanisms underlying this control are poorly understood. The main difficulties in the elucidation of these mechanisms stem from an incomplete understanding of:

1. The dynamic and static transduction properties of the sensory receptors
2. Central integration of the discharge pattern from a specific functional group of afferents
3. Central interaction among various afferent inputs
4. The end organ response to the motoneuron output

One of the least understood areas in reflex control and regulation is that of the central integration of the afferent input. The discovery process used to identify dynamic relationships between afferent input and efferent output requires precise knowledge of the input and output variables. Under conditions in which the natural (adequate) stimulus to a functional group of receptors has been systematically varied, the resulting afferent pattern to the CNS is not precisely known. Some of the reasons for this are

1. The stimulus cannot be isolated to a specific functional group of afferents.
2. The magnitude of the stimulus at the receptor site is not known.
3. The sensitivity, threshold, and saturation properties of the receptors are, themselves, reflexly regulated.
4. The receptors may respond to more than one type of physical quantity.

On the other hand, electrical activation of afferents has the advantage of generating a variety of precise input patterns which can be synchronized with the pertinent physiological events. However, its usefulness is limited to the degree with which it is capable of specific activation of one functional type of afferent input. In a nerve with more than one type of functional afferent fibers, the specificity of electrical activation is related to the degree of overlap in fiber diameters among the functional groups. The ideal situation in a mixed nerve is one in which there is sufficient separation of fiber diameters among functional groups, and the functional group of interest has the largest diameters in the nerve. This situation is approximated in the vagus nerve where the slowly adapting pulmonary stretch receptors (PSRs) appear to have the largest diameter fibers.[13-15] Thus, it is the goal of this chapter to describe a method, in which electrical activation of a functional group of afferents in conjunction with a model based on neurophysiological principles and phenomena and its validation process has lead to a better understanding of the way in which an afferent input signal is centrally processed. The results of this study have provided additional insights in the search for neuronal correlates of the processing.[17] The specific example to be used deals with the central integration of PSR input in the control of expiration.

The timing of the inspiratory (I) phase and expiratory (E) phase of the respiratory cycle is profoundly affected by afferent input arising from PSRs.[2,10] The reflex arising from these receptors is known as the Breuer-Hering reflex after its description by these investigators in 1868.[1,9] The receptor endings are located in airway smooth muscle and their fibers are located in the vagus nerves. The PSR discharge frequency has a quasilinear relationship with transpulmonary pressure.[12] These afferents make synaptic connections with neurons located in the respiratory centers of the medulla and pons where the respiratory oscillator is located.[3]

During normal breathing (eupnea), inspiration is mediated primarily through the contraction of the diaphragm, while expiration results from passive relaxation of the diaphragm and elastic recoil of the lungs and chest wall. The phrenic nerve contains lower motoneurons which innervate the diaphragm. The phrenic motoneurons receive their major input from the bulbospinal neurons which have an augmenting ramp discharge pattern during I and are silent during E. Thus, the phrenic discharge pattern, itself, is a good index of respiratory center output[6] and is useful for determining T_I and T_E. The slope of the ramp increases with an increase in the level of chemoreceptive input.

An increase in PSR input during the I-phase, as occurs during normal lung inflation, terminates the I-phase earlier with little or no effect on ramp slope.[3] Thus, tidal volume decreases with the shortening of I. An increase in PSR activity during the E-phase results in an increase in T_E and a reduction in breathing frequency. In addition, central mechanisms exist which give rise to a linear relationship between T_I and the following T_E,[2,4] as well as a (1) quasilinear relationship (2) T_E and the following T_I.[7,18] Hence, an additional advantage of electrical activation of PSR afferents is the ability to isolate the discharge pattern exclusively to one phase or the other without overlap.

It is evident from the previous discussion that the Breuer-Hering reflex offers some very distinct advantages in the study of the input/output relationships of a neural system. These include

1. Anatomic separation of the afferent and efferent limb
2. Specificity of afferent activation with electrical stimuli
3. Strong, stereotyped, reproducible responses
4. Constant ventilation and blood gases without reflex alteration in central breathing pattern as a result of bilateral vagotomy

II. APPROACH

The rationale behind the modeling process is to clarify our thinking about a problem by explicitly identifying and clearly stating every assumption and limitation. The ultimate objective is always to obtain observable implications which can be compared with experimental results.[8] Models provide a mechanism for rigorously exploring the observable implications of physiological hypotheses and thus can help to design experiments to test them and to either accept or refute the applicability of certain hypotheses in a given system under specific conditions.

The acceptance of a model is based on its ability to describe all the observable phenomena relative to the system under study. Thus, it is paramount to design and execute experiments

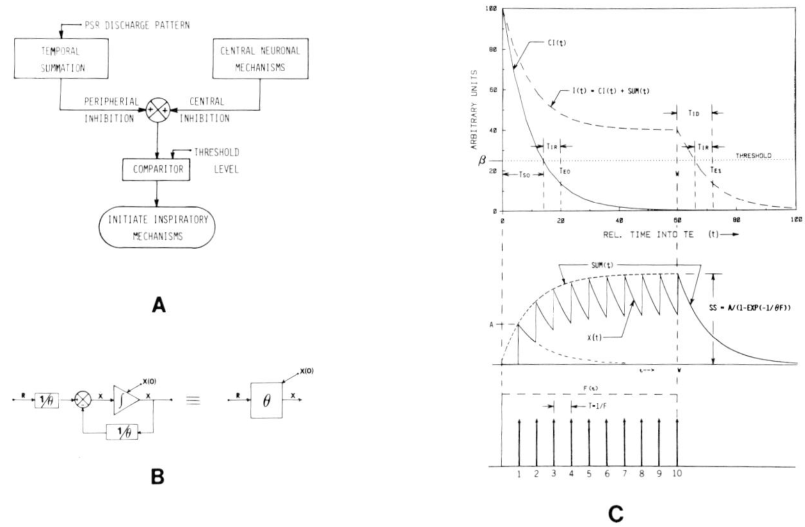

FIGURE 1. (A) Block diagram depicting the hypothesized mechanisms which underlie the determination of T_E. (B) System diagram of an integrator with negative feedback which is used to simulate the kinetics of the hypothesized processes. (C) Graphical diagram of hypothesized mechanisms for the determination of T_E. See text for additional description. (Part C, from Zuperku, E. J., Hopp, F. A., and Kampine, J. P., *J. Appl. Physiol.*, 52(5), 1296, 1982. With permission.)

which provide reliable quantitative data describing the relationships among variables. In dealing with a complex system, it is desirable to maintain all input variables constant, while systematically manipulating the one input variable of interest and observing the resulting output response. This is best accomplished under open loop conditions. In this case the PSR afferent discharge pattern is the input and T_E, as obtained from the phrenic discharge, is the output.

The ultimate goal of the studies, which are to be described, was to obtain a model which could predict the value of T_E in response to an arbitrary input pattern from PSRs. The dynamic processes described by the model provide insight into the manner in which the afferent signal is processed, the nature of the central phase timing mechanisms, and the possible neurophysiological correlates underlying phase timing and control.

III. SYSTEM MODEL

A. Hypotheses and Description

Evidence for the hypothesized model processes is based on observable phenomena and neurophysiological principles. The details of this evidence are presented elsewhere.[18] The initial assumptions are summarized as follows:

1. In the absence of PSR input, there exists a central inspiratory inhibitory (CII) process which is maximum at the beginning of E and decays, exponentially with time into E. This is shown diagramatically in Figure 1A and graphically in Figure 1C, upper.
2. There is a threshold level, β, which the CII function, CI(t), crosses (Figure 1C). After threshold crossing, there is a short delay before the onset of inspiration, during which time central inspiratory mechanisms appear to be in the process of enabling inspiratory neuron discharge (Figure 1A and C).
3. The delay period, Tir, is considered to be independent of the PSR discharge and is treated as a simple delay.

4. The PSR discharge is assumed to undergo temporal summation (leaky integration). The temporal summation of a pulse train is depicted in Figure 1C, center. At the beginning of E, the summation amplitude is assumed to be zero, since the PSR activity during I appears to contribute to CII. The PSR activity during I appears to be "gated" to the functional I centers and results in a shortening of both I and E, whereas activity during E, lengthens E, and the T_E of the previous breath has no effect on the T_E of the present breath.[18]
5. The total inhibition to inspiration, I(t), is assumed to result from the sum of CI(t) and the peripheral or vagal inhibition, SUM(t), i.e., I(t) = CI(t) + SUM(t), where SUM(t) is the envelope of the peak values of the summation, x(t). T_E is then determined when I(t) crosses threshold. This is depicted in Figure 1C (top) where a pulse train of width, W, produces a $T_E = T_{E1}$.

These initial assumptions will be tested by observing how closely this model is able to predict the relationships between T_E and the various PSR discharge patterns obtained experimentally.

B. Mathematical Representation

The basic elements of the model are a temporal summer that processes the vagal input, an exponentially decaying central component, and a threshold, which, when crossed by the sum of the central and processed vagal component, terminates the E-phase.

A diagram of a temporal summer is shown in Figure 1B. It consists of an integrator with negative feedback, where R is the input forcing function, θ is the time constant, x is the output, and $\dot{x}$($\dot{x}$ = dx/dt) is the input to the integrator. The input gain block, $1/\theta$, is used to allow the output response amplitude to be independent of the time constant, and x(0) is the value of x at t = 0 (initial condition). The system is described by:

$$\dot{x} = (-1/\theta) \cdot x + (1/\theta) \cdot R \tag{1}$$

Equation 1 is a first-order linear differential equation with constant coefficients. The presence of the forcing function, R(t), requires a particular solution, as well as the complimentary solution for x(t).

The impulse response of a temporal summer is given by:

$$x(t) = A \exp[-(t - t_o)/\theta] \tag{2}$$

where A is the response amplitude, θ is the time constant, and t_o is the time at which the impulse occurs. This response is shown by the decaying dashed line in Figure 1C, center. An impulse into this system is equivalent to increasing the value of x at $t = t_o$ by the amplitude, A, of the impulse. Thus, the initial condition on the integrator is increased by A.

It can be demonstrated that the envelope, SUM(t), of the temporal summation, x(t), of a step discharge pattern can be obtained by applying a continuous step to the input (see also Appendix B of Reference 18). The steady-state amplitude, SS, of this step response is a nonlinear function of frequency and has the form:

$$SS = A/[1 - \exp(-1/F \cdot \theta)] \tag{3}$$

where F is the discharge frequency. The envelope of the temporal summation function is

$$SUM(t) = SS \cdot [1 - \exp(-t/\theta)] \tag{4}$$

When time constants greater than 0.5 sec are considered, as suggested by vagal reflex studies, SS can be approximated as a linear function of F for F > 5 Hz. This relationship is

$$SS = A \cdot \theta \cdot F \quad (5)$$

and the frequency requirement easily falls within the present physiological range for PSRs.

The implications of Equations 4 and 5 provide for a simplification in the analysis which allows a series of impulse intervals to be replaced by the waveform of their discharge frequency. The reflex gain, γ, provides the necessary proportionality constant which relates the steady-state amplitude, SS, to F of a step input (i.e., $\gamma = SS/F$).

The central decaying component, CI(t) (Figure 1C, top), can be generated by applying an initial condition of $x(0) = 1$ to the output of a temporal summer at the beginning of E. The output response is given by Equation 2 with $t_o = 0$.

The total inhibition, I(t) (Figure 1C, top), is given by:

$$I(t) = CI(t) + SUM(t) \quad (6)$$

T_E is determined when $I(t) = \beta$.

Details for the computer simulation of the functions in Equation 6 are given elsewhere.[18] To simplify diagrams, the block "θ" at the right of Figure 1B is used to represent a normalized temporal summer.

It must be emphasized that the validation of the model critically depends upon reproducible, quantitative relationships between the input pattern and the value of T_E. These relationships can be meaningful only when care has been exercised in selecting specific stimulation parameters and all other inputs to the control system are maintained constant.

IV. EXPERIMENTAL METHODS

A. Physiological Preparation

These studies were carried out in anesthetized (pentobarbital sodium, 30 mg/kg), paralyzed, artifically ventilated, mongrel dogs. More specific details of these procedures are presented elsewhere.[17] Both vagi were sectioned so as to allow a synthetic afferent pattern to be substituted for the natural one. An added advantage is that lung stretch during artificial ventilation has no reflex effect on the phrenic output pattern and, thus, blood gases can be maintained constant. The moving-time average of the phrenic discharge, PNG(t), was used as an index of the respiratory center output. Timing pulses derived from the upstroke and peak of the PNG(t) were used to obtain on-line values of T_I and T_E and to synchronize the stimulation system.

B. Neurophysiological Procedures

Phrenic nerve activity was obtained from the central end of the C5 rootlet. This nerve was desheathed, placed on a pair of fine hook electrodes, and submersed in a pool of warm mineral oil. A low-noise, high-input impedence, AC preamplifier in conjunction with a band-pass filter-amplifier (−3 dB at 100 and 1000 Hz) was used to amplify the phrenic activity. The moving time-average of the phrenic activity, PNG(t), was obtained by precision full-wave rectification and low-pass filtering (effective time constant, 100 msec; 4th-order filter; 1/2 amplitude at 3.5 Hz). A record of the phrenic nerve discharge and its corresponding PNG(t) are shown in Figure 2. The figure also shows the definition of T_I and T_E. The peak amplitude of PNG(t) at the end of I (V_{Tes}) is linearly related to tidal volume.[5] A comparitor and monostable multivibrator were used to generate a timing pulse at the upstroke of the

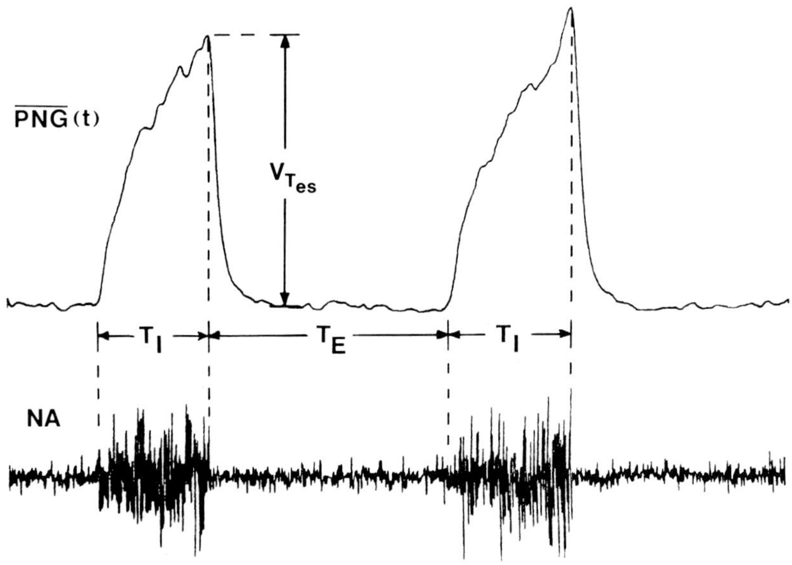

FIGURE 2. Measurement of central respiratory phase timing; lower, record of multifiber nerve activity (NA) from the C5 rootlet of the phrenic nerve; upper, phrenic neurogram or moving time-average, PNG(t), of the phrenic nerve activity shown in the lower recording; V_{Tes}, estimated tidal volume.

PNG(t). The rapid falling slope immediately following the peak of the PNG(t), which was obtained from the derivative of the PNG(t) signal, was used to trigger a comparitor and monostable multivibrator and generate a timing pulse at the end of I. Values of T_I and T_E were computed on-line by digital timers with digital-to-analog output. Both cervical vagus nerves were isolated and sectioned caudally. Approximately 6 cm of the sheath were removed from the central end of the left vagus nerve. The desheathed nerve was placed on silver hook stimulating and recording electrodes and submersed under warm mineral oil. The vagal compound action potential was used to monitor the conduction velocities of the stimulated fibers. A battery-powered, optically isolated, constant-current output unit was used to deliver rectangular pulses to the vagus nerve. We have found that it is important to desheath the nerve in order to obtain a very consistant stimulus over long periods of time (6 to 8 hr).

V. ELECTRICAL STIMULATION: TECHNIQUES AND JUSTIFICATION

A. Vagal Stimulation Procedure

To prevent any type of accumulative effect due to prior input, it is desirable to separate test breath stimulus patterns with a number of control breath patterns. In addition, for each test breath, it is desirable to randomize the presentation of the test pattern parameters. To achieve this goal, a stimulation system was designed and fabricated to apply a predetermined temporal pattern to each breath.[9a] This stimulation system consists of a calibrated voltage-to-frequency convertor (VFC nonlinearity, 0.5% maximum; frequency range, 1 to 200 Hz), calibrated waveform generators synchronized with the PNG(t) timing pulses, analog selector switches, a digital programmable control unit, and an isolated constant-current pulse-output unit (Figure 3). The control unit was programmed to select one of four possible waveforms by means of the analog switches for a particular breath. The address register was incremented at the time of the upstroke of the PNG(t), and information in the corresponding memory register was used to close the appropriate switch. This resulted in presenting the proper waveform to the input of the VFC. Any sequence of patterns could be programmed (maximum

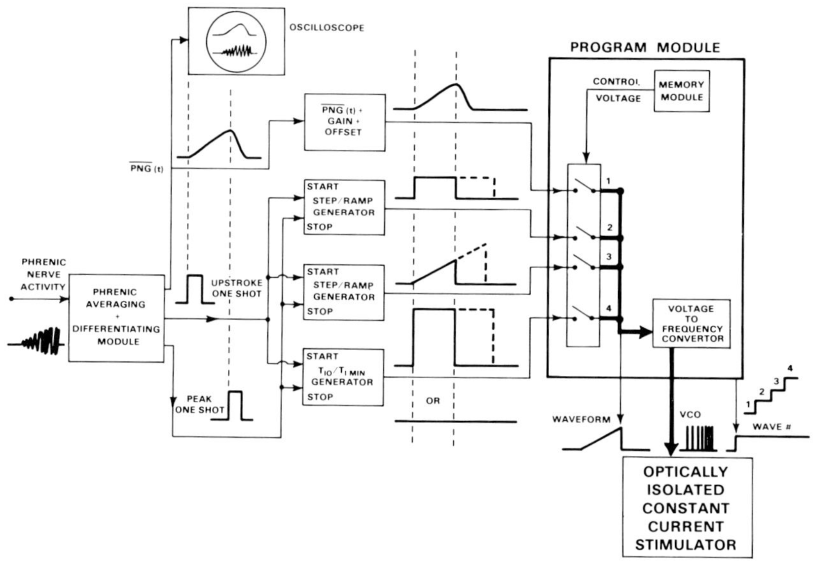

FIGURE 3. System diagram and waveforms of the programmable stimulator. See text for explanation. VCO = VFC. (From Hopp, F. A., Zuperku, E. J., and Kampine, J. P., *J. Appl. Physiol.*, 54(4), 1149, 1983. With permission.)

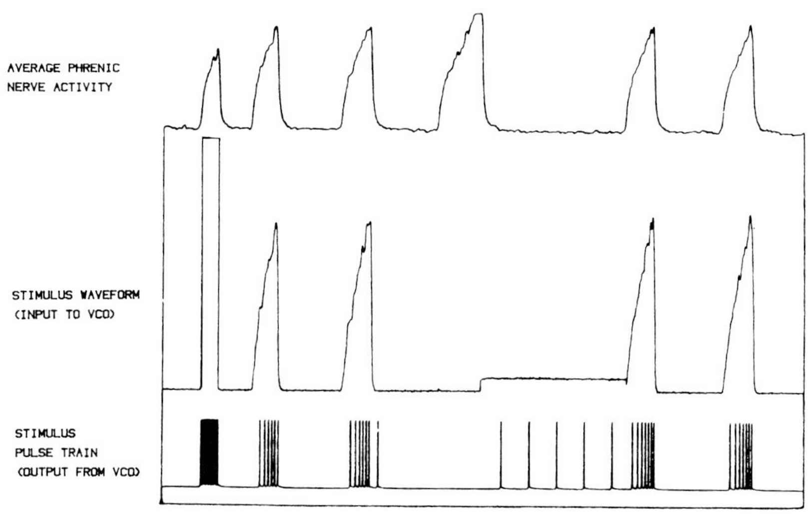

FIGURE 4. Recordings depicting the effects of various stimulus patterns on phrenic activity and respiratory phase durations; upper, phrenic neurogram; middle, stimulus frequency waveform; lower, stimulus pulse patterns which correspond to the frequency waveforms (middle). See text for further explanation. VCO = VFC.

program of 256 steps). An instruction was available to allow recycling through the same program as often as desirable. The waveform generators were capable of generating patterns consisting of calibrated steps, ramps, pulses, and delayed pulses. Any of the patterns could be terminated by the PNG(t) timing pulses. Accurate pulse-train delay times and pulse-train widths were generated by a crystal controlled time base and digital timers. For a given test waveform, the parameters of that waveform were manually controlled from the front panel of the corresponding waveform generator.

An example of the stimulation system output and its reflex effect on the PNG(t) is shown in Figure 4. In the first breath of the record, a high-frequency step input was applied during

I only and resulted in shortening T_I. During the second and third breath, the PNG(t) was used as input to the VFC. The amplitude and baseline of the PNG(t) were adjusted (via signal conditioner) to produce values of T_I and T_E similar to their prevagotomized values. The time-course of the control pattern mimics that of the PSRs during inflation. During the fourth breath, no input pattern was applied during I, resulting in a reflex lengthening of I. A step pattern was applied during the E-phase of the fourth breath, resulting in a reflex increase in T_E. The control pattern was again applied to the fifth and sixth breaths.

B. Selection of Stimulus Strength

The purpose of these studies was to focus on the role of the PSRs in the control of T_E. Since there exists a considerable overlap in the conduction velocity histograms of the vagal fibers carrying impulses from PSRs and ''irritant'' or rapidly adapting pulmonary receptors and, perhaps, nonpulmonary receptors,[13-15] a procedure was adopted to select the current strength for the optimal expiratory facilitatory reflex. The results of such a procedure are illustrated in Figure 5 (top), where the ratio of T_E to its nonstimulated value, T_{EO}, is plotted as a function of the current strength, Is, for step patterns throughout E (step frequency = 20 Hz). The pulse duration was maintained at 100 μsec. As the number of depolarized afferent fibers increases with an increase in current strength, T_E increases, reaches a maximum value, and then decreases. The current strength of maximum effect appears to be independent of the step frequency.[17]

The vagal compound action potentials obtained for stimulus strengths of 125, 150, and 170 μA are shown in the insert of Figure 5, bottom. It can be noted that it is not possible to determine the current strength at which the maximum effect occurs from these potentials alone. The conduction velocities were typically in the range of 40 to 60 m/sec and are indicative of large mylinated afferents.

These results suggest that the E facilitatory reflex appears to be antagonized as the current strength recruits more of the smaller-diameter fibers, perhaps from the irritant receptors. For the same current strengths, step inputs during I shorten T_I monotonically as strength increases (Figure 5, bottom).

The current strength used in each of these studies was determined from the maximum T_E response to a step input during E. Once determined, the strength was maintained constant throughout the various stimulus protocols used. If for some reason the stimulus electrodes were disturbed, this procedure is repeated. Only those relationships obtained at the same current strength were compared in the same animal.

C. Natural vs. Electrical Activation of Afferents

With current strengths selected according to above described procedure, it was possible to mimic the E facilitatory reflex mediated by the PSRs. This was demonstrated in anesthetized, open-chest, paralyzed dogs on artificial ventilation. A section of each intact cervical vagus nerve was desheathed, placed on stimulating electrodes, and submerged under warm mineral oil. A solenoid ventilator was triggered at the upstroke of the PNG(t). Constant inspiratory airflow was permitted during I only. The resulting tidal volume for each control breath was the product of flow and T_I for that breath. After every nine control breaths, the ventilator connected the lungs to a constant-pressure reservoir through a separate solenoid valve. This valve remained open until the onset of the following breath. The E facilitatory reflex can be seen in Figure 6, left.

Following long T_E values, 30 to 40 control breaths were used to allow recovery to a stable breathing and expired CO_2($ETCO_2$). The values of constant pressure for test breaths were randomized (range 0 to 5 torr).

After sufficient mechanical data were obtained, the vagi were sectioned bilaterally, and the central ends were electrically stimulated. The transpulmonary pressure waveform (P_t,

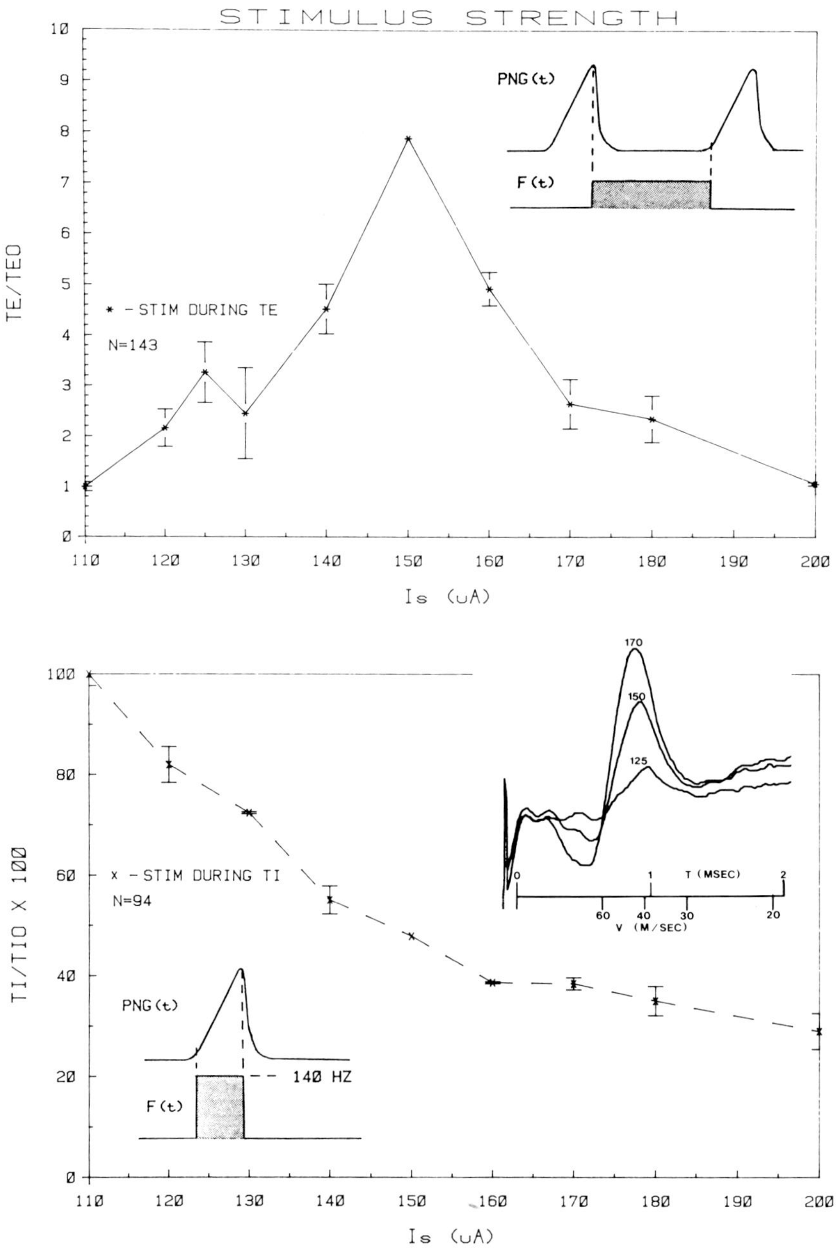

FIGURE 5. Upper panel: plot of T_E/T_{EO} vs. stimulus current amplitude (Is, μA). Stimulus pulse width was maintained at 100 μsec. A step pattern of 20 Hz was applied throughout E (insert, top). Is < 110 μA had no effect. Time drift in T_E was minimized by normalizing each test breath T_E value, T_E(Is), with respect to its nearest T_E value obtained at Is $= 150$ μA, $T_E(150)$. The normalized T_E is given as: $T_E^*(\text{Is}) = [T_E(\text{Is})/T_E(150)]\cdot\bar{T}_E(150)$, where $\bar{T}_E(150)$ is the mean of all $T_E(150)$ values. Thus, no SE bars appear at the 150-μA value. $T_{EO} = 6.73$ sec. Lower panel: Plot of T_I/T_{IO} vs. Is for step patterns throughout I (lower left insert); step frequency, 140 Hz. Time drift normalized as described above using $T_I(150)$. $T_{IO} = 1.85$ sec. Insert, upper right: averaged vagal compound action potential at Is $=$ 125, 150, and 170 μA (32 averages). Horizontal scale: time from trailing edge of stimulus, T (msec), and conduction velocity, V (m/sec).

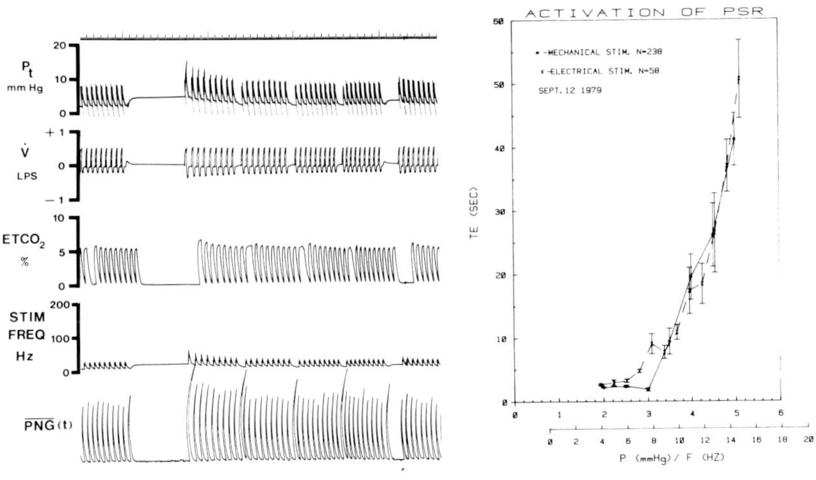

FIGURE 6. Comparison of reflex changes in T_E which result from natural of PSR discharge (prevagotomy) with those which result from electrical activation of PSR inputs (postvagotomy). Left panel: polygraph records of (from top to bottom) tracheal pressure (open chest), P_T, airflow, $\dot{V}$, airway CO_2 concentration $ETCO_2$, stimulus frequency pattern (input to VFC), and the phrenic neurogram, PNG(t). Test breaths consisted of static pressures throughout one breath. Stimulator was turned off during the prevagotomy data collection. After bilateral vagotomy, the stimulus pattern was derived from P_T, and the same amplitudes of static lung inflation were repeated. Right panel, plot of T_E vs. static pressure (solid line; upper abscissa scale, 0 to 6 mmHg) and plot of T_E vs. static stimulation frequency (dashed line; lower abscissa scale, 0 to 20 Hz). Plots are superimposed. Relationship between frequency and pressure: $F = 3.5 \cdot P - 2.7$. N is the number of test breaths used in the plots.

Figure 6) was used to generate the stimulus pattern in conjunction with the VFC.[16] The tonic frequency of the stimulator was adjusted to give the same T_E values as those observed at zero airway pressure. The gain of the frequency waveform was adjusted to give approximately the same T_E values observed at 4-torr airway pressure (prevagotomy).

The same lung inflation maneuvers were then repeated with electrical activation of vagal afferents substituted for mechanical activation. A plot of T_E (test breaths) vs. inflation pressure (prevagotomy) is superimposed on a plot of T_E vs. stimulus frequency (postvagotomy, Figure 6, right). The close agreement of these results indicates that the electrical activation of vagal afferents can, rather faithfully, reproduce the reflex changes in T_E that arise from mechanical activation of PSRs.

VI. MODEL VALIDATION

The validation process is the primary operation which leads to an acceptable model. It is basically part of a discovery process which eliminates those model assumptions that are incapable of accurate prediction of the experimently observed phenomena.

The steps of the validation procedure are outlined in Figure 7A. It is the intent of the diagram to give equal weight to the importance of both the empirical relationships and the hypothesized model. The model represents a deterministic system in that for each input there is a unique output. The empirical data used for validation was the mean output value, T_E, derived from several responses to the same input test pattern. In these studies the variability of individual T_E values appears to be large, whereas the SD of the mean values (SE) is much smaller, indicating the consistancy of the experimental conditions and the small amount of time drift of other contributing factors.

For a given plot and a given set of model assumptions, computer simulation was used to arrive at a set of parameter values, by trial and error, which minimized the differences

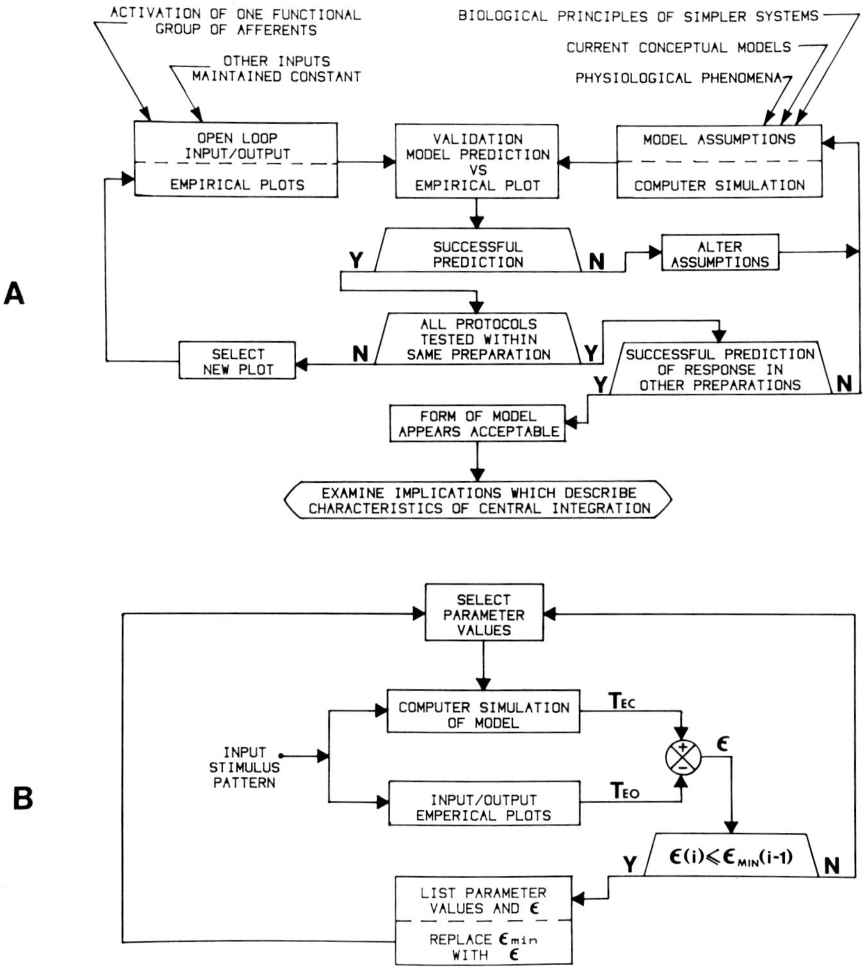

FIGURE 7. (A) Flow chart of model validation procedure; (B) flow chart for estimation of model parameters based on least squared error.

between observed and computed values. This process is depicted in Figure 7B. A set of model parameters was selected, and for a given plot such as T_E vs. step frequency, each step frequency which was used to obtain the experimental data was presented as input to the model. The error, ϵ, represents the difference in the observed and computed T_E values (i.e., $\epsilon = T_{EO} - T_{EC}$). This error was squared and summed for each of the input frequencies used. The error, for the ith set of parameters, was compared with the previous minimum error. If the ith error was less, then a list of the parameter values and the corresponding error was published, and the previous minimum was replaced by the latest minimum error. A new set of parameter values was then selected and the process repeated. This process was halted after all of the combinations of parameter values had been examined. At first a wide range of parameter values was covered with coarse increments in value. The ranges of values producing the relative minimums were then further examined with refined increments.

Those parameters which resulted in the minimum error were used to generate a curve which was superimposed on the empirical plot. An example of this is shown in Figure 8a which is a plot of T_E vs. step frequency. The insert is a block diagram representing the original model assumptions. The time constant of the CII is θ_1, and that of the temporal summation process is θ_2; the two components are additive. The experimental data are shown

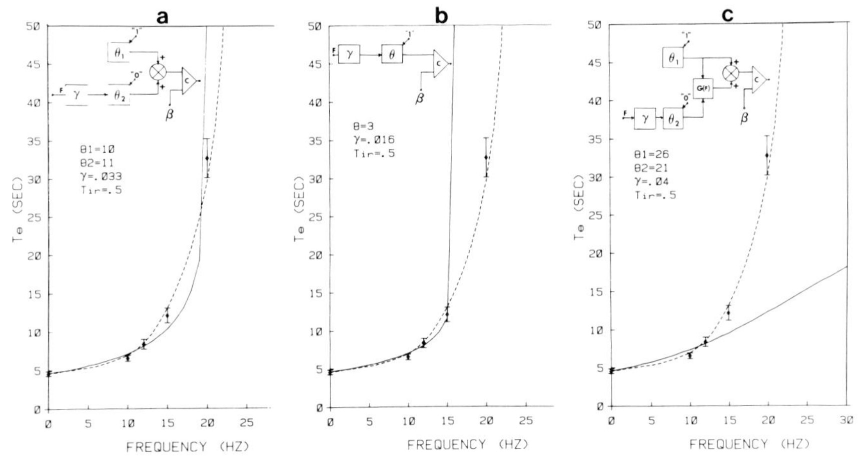

FIGURE 8. Comparison of the output (solid curves) of various model configurations with an experimental plot of T_E vs. step frequency (mean $T_E \pm$ SE). Dashed curve represents output of the "best" model (described later) and serves as a reference demonstrating a "good" fit. Insets are functional block diagrams of models evaluated and their best fit parameters. See text for further details. (From Zuperku, E. J., Hopp, F. A., and Kampine, K. P., *J. Appl. Physiol.*, 52(5), 1296, 1982. With permission.)

as points with ($\pm$1) SE bars. The solid curve is the model output "best fit" and the corresponding parameter values are listed in the insert. The dashed line is used for the sake of comparison and was obtained from the model (to be discussed later) best able to predict T_E values for a number of arbitrary discharge patterns. From the figure it can be seen that this model output is not able to predict the larger values of T_E with sufficient accuracy. This inability was even more obvious for the same type of plot from other animal preparations.

The model assumptions were altered (upper right-hand loop of Figure 7A) by the constraint that $\theta_1 = \theta_2$. This arrangement implies that the central and vagal components are processed by the same functional operator and that there is a common mechanism for inhibition. The block diagram of this model is shown in the inset of Figure 8b. There is very good agreement between T_E values for the frequency range of 0 to 15 Hz; however, this model is not able to predict T_E values at the higher step frequencies, because the function is much to steep.

The model assumptions were altered to provide for multiplicative interaction between the central and vagal components, as opposed to summative interaction. The modulation of the CII had the form: $I(t) = [1 + G(F)]*CI(t)$, where G(F) is the temporal summation of F(t). When $F = 0$, $G = 0$, and $I(t) = CI(t)$ as in the previous model. The output of this model is shown by the solid curve in Figure 8c. The block diagram and best-fit parameters are given in the inset. Because of the exponential nature of CI(t), this model is not capable of producing a complete central apnea (i.e., $T_E \rightarrow \infty$). This assumption was discarded.

The common mechanism model of Figure 8b was then assumed to be a second-order system as depicted in the inset of Figure 9b. The output of this model is shown as the dashed lines in Figure 8 and the solid line in Figure 9a. The best-fit parameters are listed in the inset of Figure 9c. This model was capable of fitting the T_E vs. step frequency data with a high degree of accuracy.

In order that this model may not be regarded as a elegant nonlinear regression function, its ability to accurately predict responses for other input patterns was examined. For this purpose, its output was compared with T_E plots generated from other stimulus patterns taken successively from the same animal. This process is indicated by the left-hand loop in Figure 7A. The values of each of the model parameters were estimated from the best fit curves of the T_E vs. step frequency plots. These values were then used to predict the T_E values generated

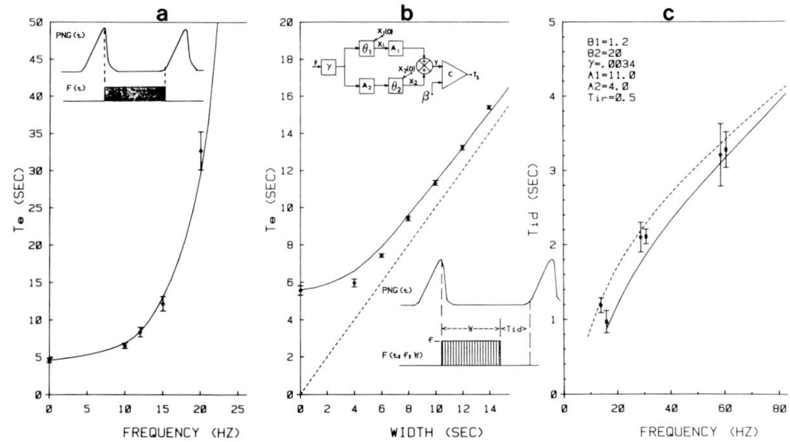

FIGURE 9. (a) Plot of T_E vs. step frequency (mean $T_E \pm$ SE). Same data are in Figure 8. Inset depicts stimulus pattern used during test breath. Solid curve is the output of the parallel two-component T_E model shown in the inset of b, upper. Data of a were used to obtain best estimates for model parameters (inset, c). (b) Plot of T_E vs. pulse train width, W (waveform, lower inset); pulse-train frequency, 15 Hz; dashed line, time of falling edge of stimulus train (i.e., T_E = W). Solid curve is model output for parameters obtained from data of a. (c) Plots of T_{Id} vs. pulse-train frequency at a train width of 8 (*) and 12 sec (0). Data and curves have been staggered for clarity. Dashed line is model output for W = 8 sec, and solid line is model output for W = 12 sec. (From Zuperku, E. J., Hopp, F. A., and Kampine, J. P., *J. Appl. Physiol.*,52(5), 1296, 1982. With permission.)

by other stimulus patterns. The time required to gather experimental data for each plot was on the order of 1 hr. Thus, due to the long period of time between the beginning of each stimulus period, a small amount of drift in the breathing pattern usually occurred. For this reason, the threshold level, β, was calculated to provide agreement between the experimental and calculated T_E values at the y-intercept of the plots. Based on the same model parameter values, good agreement between the predicted and experimental values for various stimulus patterns would provide further evidence for the validity of the model.

A second type of stimulus pattern consisted of a constant frequency pulse train in which the width, W, was increased from the onset of E. The relationship between T_E and W has a form such as that shown in Figure 9b. The data for plots a to c of Figure 9 were obtained from the same animal. The dashed line represents the time of the end of the pulse train. The plot of T_E vs. W becomes approximately parallel to the time of the falling edge of the train. The time difference between T_E and W, T_{Id}, becomes constant for values of W greater than 6 sec in Figure 9b. It appears as if the falling edge of the train is pushing back the onset of the next I by the constant amount, T_{Id}. This phenomena can be visualized in Figure 1c, top, when I(t) plateaus, such that the distance of I(t) above the threshold becomes independent of time. The output of the second-order model based on the model parameters estimated from the step frequency data (Figure 9a) is shown as the solid line in Figure 9b. There is good agreement between the data and model output for widths greater than 6 sec.

The model suggests that the time by which I is delayed from the falling edge of the train, T_{Id}, should depend on the amplitude of the plateau level relative to the threshold. Furthermore, long-time constant temporal summation suggests that the plateau level is proportionate to the discharge frequency. Plots of T_{Id} vs. stimulus frequency at train widths of 8 and 12 sec are shown in Figure 9c. Frequencies of 15, 30, and 60 Hz were used. The differences in T_{Id} at the two widths are not significantly different ($p \leqslant 0.025$). The model output based on the parameters estimated from the step frequency data (Figure 9a) is in good agreement with the experimental data (Figure 9c). The plots have been staggered for clarity.

Other stimulus patterns, which consisted of combinations of tonic and phasic patterns, were used to obtain relationships between T_E and waveform parameters. The model output was also able to satisfactorily fit these plots which were obtained from the same animal preparation.[17]

The predictive capabilities of the model were further examined with pairs of plots obtained from other animal preparations. The first plot, usually T_E vs. step frequency, was used to obtain estimates of the best-fit parameter values. These values were then used to see how well the model output predicted the second plot. In 16 out of 18 plots, the average error (difference between computed and observed values) was less than 10%.[17] These results provide additional support for the validity of the model.

VII. MODEL PROPERTIES

A. Uniqueness of the Minimal Realization

The results of these studies suggest that the PSR input is processed by slow processes (approximately $\theta_1 = 0.8$ sec, $\theta_2 = 18$ sec) and that the central I inhibition is comprised of the same slow processes. Variations of the model components were also examined. The case in which the central component was assumed to be a single decaying exponential, whereas the vagal component was assumed to be processed by a second-order system, was unable to provide a good fit to the T_E vs. step frequency plots. Likewise, the case in which the central component was assumed to be made up of two exponentials, while the vagal input was processed by a first-order system, was unable to provide good fits. Thus it appears that the central and peripheral inputs each require two components.

A rearrangement of the elements in the second-order system model allows simulation of accommodation similar to that observed in nerve and described by the Hill-Rashevsky-Monnier model.[11] The output of this model was able to provide a reasonable prediction of T_E for the T_E vs. step frequency and T_E vs. W plots of Figure 9a and b; however, it was incapable of predicting the T_{Id} vs. F relationships of Figure 9c. It was further noted that it was difficult to obtain good curve fits of other T_E vs. step frequency plots in the same animal.[17]

The high degree of uniqueness exhibited by the final model configuration (Figure 9b, inset), which represents a minimal realization of the model, suggests that this modeling-validation technique has sufficient sensitivity to discriminate between acceptable and non-acceptable hypotheses. This strengthens the credibility of the implications suggested by this model.

B. Implications

The following implications are suggested:

1. The time constants (TCs) of the CII and those which process the PSR input appear to be the same; thus, the PSR input and the source of the CII appear to be processed by a common or shared mechanism.
2. Processes with long TCs are involved with the generation of the CII. Possible sources of long TC processes include transmitter-receptor interactions (e.g., slow EPSPs or IPSPs), rebounds, and specialized neural network interaction.
3. The time-course of the CII is described by the sum of two "pure" decaying exponentials. The generation of "pure" decaying exponentials may have three sources: rapid removal of an existing input; the response to a step input, or an impulse response.
4. Long TC temporal summation provides a linear frequency-to-amplitude conversion, thus the PSR input frequency appears to have a linear representation centrally. A quasilinear relationship between PSR discharge frequency and transpulmonary pressure

(P_{tp}) coupled with long TC temporal summation may provide an analog signal format for P_{tp}.

5. In the presence of long TC processing, the temporal aspects of the input pattern are not preserved thus, correlations based on discharge patterns (e.g., cycle-triggered histograms of medullary I or E neural activity) alone may be misleading. In addition, information from the previous phase may be passed on to the following phase.

Input patterns evoked in PSR afferents may be used as a tool to extract the time-course of the CII and, thus, the effect of various interventions which affect the CII may be studied. In addition, the processed version of the PSR input (e.g., pulse train) should manifest itself in prospective candidates of the CII generators and thus aid in their identification.

VIII. SUMMARY

A method utilizing selective electrical activation of afferent fibers has been developed and demonstrated as a tool for the investigation of CNS integration. It consists of afferent input/efferent output relationships, a mathematical model of the system which is based on neurophysiological principles, and a validation process which eliminates those assumptions which do not lead to accurate prediction and, thus, aids in the discovery of the system function. The specific subject studied was the Breuer-Hering expiratory facilitatory reflex. The afferent limb consists of the slowly adapting pulmonary stretch receptors and the efferent output, expiratory duration (T_E), was obtained from the phrenic nerve discharge. Electrical stimulation of the PSR fibers in the vagus nerve was used to obtain precise temporal input patterns.

The results of this study have led to the development of a model which is capable of accurate prediction of T_E in response to arbitrary input discharge patterns from PSRs. Thus, it is felt that the dynamics of the processes it simulates are similar to the actual ones involved with E-to-I phase switching. The high degree of uniqueness exhibited by the final model configuration suggests that this modeling-validation technique is very specific in its selection of an appropriate model. The model has led to several interesting implications, the most notable of which is the role of long time-constants in afferent processing and central timing mechanisms; these are two orders of magnitude greater than those of the fast EPSPs and IPSPs. It is hoped that the characterization of the underlying processes of this reflex will provide insight in the search for the neuronal correlates, as well as provide a better understanding of the controller function itself.

REFERENCES

1. **Breuer, J.,** Die Selbltsteuerung der Athmung durch den., *Nervus Vagus, Sitzber. Akad. Wiss. Wien.*, 58, 909, 1868.
2. **Clark, F. J. and von Euler, C.,** On the regulation of depth and rate of breathing, *J. Physiol. (London)*, 222, 267, 1972.
3. **Cohen, M. I.,** Neurogenesis of respiratory rhythm in the mammal, *Physiol. Rev.*, 59, 1105, 1979.
4. **D'Angelo, E.,** Central and direct vagal dependent control of expiratory duration in anaesthetized rabbit, *Respir. Physiol.*, 34, 103, 1978.
5. **Eldridge, F. L.,** Relationship between phrenic nerve activity and ventilation, *Am. J. Physiol.*, 221(2), 535, 1971.
6. **von Euler, C.,** The functional organization of the respiratory phase-switching mechanisms, *Fed. Proc. Fed. Am. Soc. Exp. Biol.*, 36, 2375, 1977.
7. **Feldman, J. L. and Gautier, J.,** Interaction of pulmonary afferents and pneumotaxic center in control of respiratory pattern in cats, *J. Neurophysiol.*, 39, 31, 1976.

8. **Grodins, F. S,** Models, in *Regulation of Breathing, Part II*, Vol. 17, Hornbein, T. G., Ed., Marcel Dekker, New York, 1981, chap. 22.
9. **Hering, E.,** Die Selbsteuerung der Athmung durch den *Nervus Vagus, Sitzber. Akad. Wiss. Wien.*, 57, 672, 1868.
9a. **Hopp, F. A., Zuperku, E. J., and Kampine, J. P.,** A programmable electrical stimulator suitable for control of respiration studies, *J. Appl. Physiol.*, 54(4), 1149, 1983.
10. **Knox, C. K.,** Characteristics of inflation and deflation reflexes during expiration in the cat, *J. Neurophysiol.*, 36, 284, 1973.
11. **MacGregor, R. J. and Lewis, E. R.,** *Neural Modeling*, Plenum Press, New York, 1977, 159.
12. **Miserocche, G. and Sant'Ambrogio, G.,** Responses of pulmonary stretch receptors to static pressure inflations, *Respir. Physiol.*, 21, 77, 1974.
13. **Paintal, A. S.,** Vagal sensor receptors and their reflex effects, *Physiol. Rev.*, 53, 159, 1973.
14. **Paintal, A. S.,** The nature and effects of sensory inputs into the respiratory centers, *Fed. Proc. Fed. Am. Soc. Exp. Biol.*, 36, 2428, 1977.
15. **Sampson, S. R. and Vidruk, E. H.,** Properties of "irritant" receptors in canine lung, *Respir. Physiol.*, 25, 9, 1975.
16. **Trenchard, D.,** Role of pulmonary stretch receptors during breathing in rabbits, cats and dogs, *Respir. Physiol.*, 29, 231, 1977.
17. **Zuperku, E. J., Hopp, F. A., and Kampine, J. P.,** Central integration of pulmonary stretch receptor input in the control of expiration, *J. Appl Physiol.*, 52(5), 1296, 1982.
18. **Zuperku, E. J. and Hopp, F. A.,** On the relation between expiratory duration and the subsequent inspiratory duration, *J. Appl. Physiol.*, in press, 1985.

Chapter 7

ELECTRICAL STIMULATION OF THE SPINAL CORD IN MOVEMENT DISORDERS

Arthur M. Sherwood

TABLE OF CONTENTS

I. SUMMARY

Spinal cord stimulation (SCS) is an example of the use of sensory or afferent stimulation to modify abnormal motor control in patients with upper motor neuron disorders. SCS evolved from the use of stimulation to control pain and is now used by more than 30 groups around the world. Although the precise mechanisms of action of SCS are not known, a common feature of SCS is that the stimulation only modifies motor control in those patients with abnormal motor control. This is accomplished by epidural stimulation of the dorsal side of the spinal cord. The effects of SCS are presumably mediated by the dorsal columns, with secondary effects mediated by the autonomic nervous system, by additional ascending and descending tracts, by the propriospinal interneuron system, and by the gray matter of the spinal cord. When the system fails to produce the desired result, it may be due to a technical failure of the stimulation itself (e.g., improper location of the electrodes, wire breakage, or failure of electronic components); to the absence of necessary physiological mechanisms to mediate the response; or to the undesirable nature of the effects induced by stimulation. Further clinical and neurophysiological studies will be needed to establish the relative contribution of stimulation of the dorsal column vs. the contribution of stimulation of other structures. Such knowledge will facilitate the expansion of the application of this approach to other patient categories and degrees of disability.

II. INTRODUCTION

Stimulation of the dorsal portion of the spinal cord through electrodes in the epidural space has been used as a means of modifying spasticity, impaired coordination, and bladder dysfunction in patients with upper motor neuron lesions by an increasing number of workers since 1972. In a review of 22 papers representing 14 different active groups, Siegfried et al. found reports of application in 796 patients prior to March 1980.[131] SCS has been shown to result in definite, though modest, benefits in 35% to as many as 75% of well-selected patients with a variety of disorders,[131] although by far the majority of patients so treated have been those with multiple sclerosis (MS). The stimulation is unnatural in that it does not exactly mimic any natural form of sensory input and has its effect in the partial transformation of the preexisting abnormal control. Benefits to be expected from such stimulation are dependent on the nature of the motor disorder, but are primarily in the nature of increased endurance, better bladder control, and reduced muscle hypertonia. While this field is a very active one, with a number of groups involved in research and clinical studies, there are numerous clinical and technical problems as well. The modest nature of the changes effected by SCS in most cases has made it difficult to quantify changes induced by SCS, even when patients report that the changes have been significant in modifying their quality of life. The major theoretical problem in SCS is that the underlying mechanisms are not known, making acceptance of its therapeutic application slow and progress in new methodologies difficult. Similarly, the application of SCS has been hindered by the difficulty in selecting those patients most likely to benefit from the stimulation. Much work remains to be done, not only clinically, but also in research into mechanisms and in development of new devices.

III. HISTORICAL REVIEW

A. From Dorsal Column Stimulation (DCS) for Pain Suppression to SCS for Modification of Spasticity

SCS has been used as a means of treating intractable pain since 1967, when it was introduced by Shealy et al.[177] as an application of the Melzack and Wall gate control theory of pain.[172] It has continued to be used by a number of workers.[169] While treating an MS patient for pain with SCS, Cook made the observation that such stimulation improved the patient's voluntary motor control and sensory appreciation.[12] He investigated the significance of this discovery by applying SCS in four additional patients.[12-14] He found that stimulation produced a "significant modification of abnormal neurologic signs and dysfunction in patients with M.S.". Even in this original series, there was not 100% success, as the second patient received essentially no beneficial results[12] and 1 of the 5 quit using the stimulation within 2 months due to psychological factors. Only the original patient showed more than a modest improvement in total disability.[12-14] Cook's initial report and later work induced a number of other clinicians to apply SCS in patients with MS since it offered virtually the only possibility of alteration of the patient's status.[78,92,125] Results of these and other studies have shown that while SCS does not alter the course of MS, it does provide a degree of improvement of function. Cook and others also began evaluation of the effects of SCS in other forms of upper motor neuron disorders.[25,53,66]

B. Initial Efforts: Patient Population, Techniques, and Results

SCS has been applied primarily in MS patients, but it has also been used in those with spino-cerebellar disorders, cerebral palsy, and incomplete post-traumatic myelopathy.[30] In general, patients with upper motor neuron dysfunction but with partial preservation of the function of sensory and motor pathways may be candidates for SCS.

The equipment and techniques for SCS were developed for the modification of benign, intractable pain, and indeed, such applications today comprise the bulk of the market for these devices. Initially, stimulation was accomplished by means of a laminectomy, allowing access to the spinal cord (SC) and its covering, the dura. The dura was surgically opened and an electrode array embedded in silastic mesh was implanted in the extra-arachnoid space over the center of the upper thoracic SC.[140] Electrodes were activated by an implanted receiver powered and controlled by an external transmitter through a radio frequency (RF) transformer coupling implemented by placing an antenna on the skin surface over the receiver. Stimulation was usually continuous, at least during the day, and was adjusted to a level producing tingling or paresthesias in the trunk, back, and possibly legs as well. SCS for motor control has tended to employ a standardized placement of stimulating electrodes, the choice of which depended on the type of motor disorder being treated.

Evaluation of the effectiveness of SCS for motor disorders has been based primarily on clinical observations, with some exceptions.[40,103,131] The results reported include a modest reduction in spasticity, improvement in endurance for physical activity, and improvement in bladder control. Beneficial results began within a few hours in a limited number of cases, but usually within 24 to 48 hr of the onset of continuous stimulation. In many cases, patients continued to improve for as much as a month or more after the stimulation was begun, after which changes in motor function reached a plateau. However, interpretation of results has in general been confounded by a variety of technical and interpretational problems.[40,117-119] At least three groups concluded that the results of SCS were not beneficial enough for them to consider it a valid treatment.[63,71,108,143] Indeed, one of these papers[108] was reviewed and presented as the authoritative finding on SCS for treatment of motor disorders in MS.[33] The majority believe that since the procedure is nondestructive and since there are no better therapeutic alternatives for these patients, SCS should be applied in suitable candidates. As already mentioned, the beneficial effects are modest, without alteration of the underlying pathophysiology and usually without changing the category of the patient's disability. Instead, the ability of the patient to function within that category is altered. For example, if the patient can walk, his endurance is better and he can walk better, with less fatigue and less spasticity, and hence can walk further; if he is wheelchair bound, he can better utilize the wheelchair by being able to control spasticity for longer periods, to more easily transfer; if the patient had spasms at night, these will be reduced, allowing him to sleep better.

C. Literature Summary

A total of more than 140 references to papers, scientific presentations, letters and reports on SCS for motor disorders is listed in the References appended to this chapter. This represents the work of some 28 active groups of clinical researchers in various centers in Europe and the U.S. Many of those using SCS for modification of abnormal motor control have attempted to employ it without understanding the physiological mechanisms involved, without demonstrating the presence of essential neurocontrol mechanisms, and without adequate biomedical engineering support to define and provide necessary technical support. This has led, on occasion, to applications of SCS in an improper environment, either excessively optimistic or unduly pessimistic. The vast majority of the patients treated with SCS have been treated by a relatively small number of centers.[29,31,61,131,140] Reports of objective, neurophysiological evaluation procedures have come from an increasing number of investigators.[40,65,77,81,90,93,103]

IV. STIMULATION METHODS

Once the decision has been made to apply SCS in a carefully selected patient, a method for placing the electrodes in close proximity to the desired portion of the SC must be chosen. In order to provide a constant level of stimulation, not only must the stimulation system be capable of a stable output, but the electrode array must maintain a reasonably constant position relative to the SC. This task is made more difficult by the fact that the cord moves relative to the fixed bony structure of the spinal canal. Finally, relative motion of various portions can cause breakage of electrode leads and the invasion of body fluids can disrupt the system by their chemical action. The methods used to address these problems are described below, beginning with a description of the available electrodes and stimulators, followed by a description of the more commonly used techniques, and ending with a discussion of stimulation parameters, safety, and problems.

A. Patient Selection

One of the primary difficulties in the application of SCS for treatment of motor disorders lies in the selection of the appropriate candidates. Potentially, patients with any upper motor neuron disorder may be considered for SCS, including MS and SCI, but also cerebral palsy, torticollis, Strumpell's familial spastic paraplegia, Friedreich's ataxia, and dystonia. It is essential to determine if the potential gains afforded by SCS will be useful to the patient, not only from the perspective of their physiological condition, but also taking into account their psychological status. In the case of MS patients, Cook stressed the importance of not only verifying the diagnosis, but also of ruling out or dealing with other complications, such as unsuspected urinary tract infections.[12,25] He further noted the importance of assessing the probable outcome with respect to the patient's present condition, particularly in light of the patient's own view of his impairment. The patient should not be too debilitated or so slightly impaired as to render the procedure meaningless. Finally, the patient must be stable emotionally. The difficulty in selecting patients is increased because neurological observations frequently are not in agreement with reports of the effects of SCS. In other words, routine tests do not evaluate those aspects of motor control altered by SCS.

In our laboratory, we follow a protocol for evaluation which includes the following elements:

1. Confirmation of the diagnosis
2. Verification of the patient's physical condition
3. Quantification of neurological deficits
4. Neurophysiological evaluation of segmental reflexes, long-loop reflexes, postural and volitional brain influence on segmental reflexes, long ascending functions, and other sensory functions to assess the degree of preservation of underlying neural control mechanisms[11]

If the patients show signs of upper motor neuron dysfunction, but with some preserved descending control and otherwise satisfy the criteria, they may be selected for temporary placement of electrodes for a trial stimulation period. It must be recognized, however, that until the mechanisms involved in the beneficial effects of SCS are fully understood, selection criteria will be less than perfect. However, SCS is not destructive and is frequently the best alternative for the patients in spite of the uncertainties in selection.

The specific criteria used in evaluating each patient depend on the patient's disability. If, for example, a paralyzed spinal cord injury patient with a cervical lesion has ongoing spasticity in an extensor pattern with no bladder or decubitus ulcer complications, he would be a likely candidate. In summary, a potential candidate for SCS must show evidence of underlying neural control mechanisms which can be augmented through additional input to the dorsal columns, i.e., there must be signs of partially preserved descending motor control as well as a partially functioning ascending system. Since the precise mechanisms involved in generating the beneficial therapeutic effects are not known, even with careful selection, it is not possible to be completely accurate in predicting which patients will benefit. In a study of 111 MS patients with temporary SCS from 3 different center, only 37 cases were considered to be suitable candidates for permanent system implantation.[131]

B. Electrodes and Leads

The role of the electrode is to deliver current in a noninjurious manner to neural structures to be depolarized. Since it has been found that cathodal current is most effective in depolarization of neural structures, the orientation and placement of the cathode in the stimulating system is an important consideration. Indeed, with respect to depolarization of long ascending or descending tracts, it can be shown that the location of the anode is immaterial as long as the anode is several centimeters or more away from the cathode (see Section VI.). Cook et al. proposed that two electrodes in the spinal canal are unnecessary, that monofocal stimulation is quite adequate.[28] However, the experience of many others suggests that the

Table 1
COMMERCIALLY AVAILABLE SCS ELECTRODE SYSTEMS

Source			Material[a]		Dimensions	
Manufacturer	Model	Type[a]/ # poles	Electrode	Lead	Diameter (mm)	Length (cm)
Avery	E355C	C-1	Pt	Pt-Ir	1	22
	E355 (flocked)	C-1	Pt	Pt-Ir	1	22
		P-10[b]	Pt		—[b]	—[b]
Medtronic	Pisces (3482)	C-1	Pt	Pt-Ir	1	28
	Sigma (3481)	C-1	Pt	Pt-Ir	1	28
	Resume (1975)	P-4	Pt	—[b]	4	11
Cordis		C-1	Pt	Pt	1	75
	(SESCE)	C-4	Pt	Pt	1	60
Neuromed	1980J	C-4	Pt	—[c]	1	40
CSC	60	C-1	Pt	PTFE	1.3	60
Clinical Technology		P-3[b]	Pt	Tefl.	3	30

Note: Addresses for the manufacturers are Avery: Avery Laboratories, Inc., 145 Rome St., Farmingdale, N.Y. 11735; CSC: Ceramic Substrates & Components, Ltd., Farnham Common, Buckinghamshire, England; Clinical Technology: Clinical Technology, Inc., 200 West 74th St., Kansas City, Mo. 64116; Cordis: Cordis Corp., P. O. Box 525700, Miami, Fla. 33152; Medtronic: Medtronic, Inc., Neuro Division, 2199 North Pascal, St. Paul, Minn. 55113; Neuromed: Neuromed, Inc., 5000A Oakes Rd., Ft. Lauderdale, Fla. 33134.

[a] Symbols: C, catheter; P, plate; Pt, platinum; Tefl., Teflon®.
[b] Variable, custom order.
[c] Proprietary.

placement of the anode is of importance in modifying the effects observed. This is obvious when evaluating the immediate effects of SCS, such as the patterns of induced paresthesias or the distribution and threshold of induced muscle twitches. The importance of the anode location may be due to practical considerations in the precise placement of the cathode or may relate to the anatomical and physiological characteristics of the neural structures depolarized.

Regardless of the type of electrode used, the major problems with the electrode and lead system are the relative movement of the electrode and underlying neural structures, either transiently or permanently, and its durability in response to the repeated movement of the system in the spinal canal, particularly as it exits the spinal canal. Various designs have been implemented to improve the lead system, most of which are based on proven cardiac pacing technology. The electrode of choice in most situations is the single-pole catheter tip electrode described below, although the increasing availability of multipole catheter electrodes may cause them to be used more frequently in the future. Arrays of disc electrodes still offer some advantages in specific applications when SCS has been shown to be effective, yet adequate placement over the proper portion of the SC was difficult to achieve through the use of the percutaneous approach. Currently available electrodes are listed in Table 1 and are described by category below.

1. Catheter Tip

In his early percutaneous applications of SCS, Cook used a temporary cardiac catheterization electrode (Davis-Geck) percutaneously placed into the epidural space.[19,21] This electrode was constructed of twisted stainless steel wire with a platinum tip, which proved

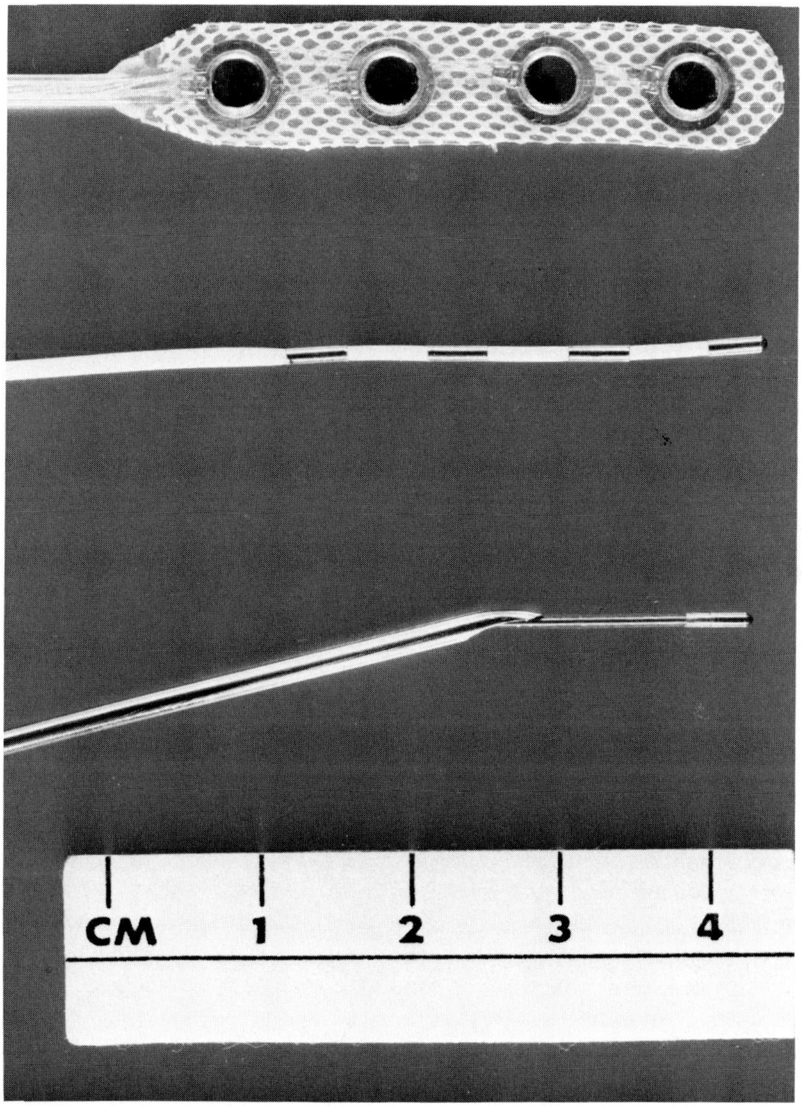

FIGURE 1. Epidural stimulating electrodes. Three commonly used electrodes are shown in this figure. The top electrode is a Medtronic Resume™ four-electrode array, designed for placement by modified laminectomy. The middle four-electrode array is the Neuromed 1980J electrode system. The bottom electrode, shown exiting the modified Tuohy needle used in placing the system, is an Avery E355 C electrode, consisting of a helical coil lead of platinum-iridium with a platinum tip. The tip diameter is 1 mm.

unsuitable due to the fragility of the lead material and due to the difficulty in placement because of the lack of stiffness. The majority of electrodes used in SCS today are composed of a thin, cylindrical tube, much like a catheter, constructed with the lead wire surrounded by insulation, extending to one or more active surfaces at the distal end (Figure 1). Most electrodes are constructed with the conductor lead wound into a helical coil shape for maximum flexibility and maximum durability with respect to imposed mechanical stresses (Table 1). The electrodes are supplied with a removable stylet inserted within the (hollow) coil in order to provide adequate stiffness and the desirable steerability to the system. The stylet is removed upon attainment of an appropriate stimulation location. The active surface of the typical catheter lead electrode system is on the order of 5 mm long and 1 mm in diameter and is constructed of platinum or platinum-iridium. A newer type catheter electrode, with four active surfaces on a single catheter, is proving to be useful in increasing the flexibility of stimulation and in attainment of appropriate stimulation sites (see Table 1).

A variety of materials have been used to insulate the electrode leads. One of the most common is silicone rubber, which is flexible, but with low tensile strength and a susceptibility to abrasion or nick-induced fracture. Teflon® has also been used in some cases. High-density polyethylene seems to have quite desirable properties in inhibiting water transmission. It has adequate tensile strength and resistance to abrasion. Another material which is suitable when properly processed is polyurethane. It has perhaps the highest tensile strength, but difficulties have been encountered in some applications due to poor manufacturing practices, which apparently destroy the hydrocarbon chain structure, causing rapid deterioration and disintegration. Properly prepared electrode leads using this material have excellent properties, however.

These flexible electrodes have a tendency to change position within the spinal canal once implanted. The Sigma® lead developed by Medtronic, Inc. and the flocked lead developed by Avery Laboratories were both designed to make the electrode lock itself in the spinal canal. The Sigma® electrode resumes a predetermined sigmoid shape over the distal portion when the stylet is removed to stabilize the system within the canal. The flocked electrode depends on the encapsulation of the flocking near the tip by fibrous tissue to provide stabilization.

2. Disc Electrodes

Several varieties of disc electrode arrays are used currently. These typically consist of a series of platinum discs fixed in a silastic mesh and arranged to be connected in a variety of ways. They are placed in position under direct visual observation following a laminectomy or partial laminectomy to expose the dura. Arrays used for pain control have also included anterior-posterior arrangements,[165] but stimulation for control of motor disorders has generally been restricted to the posterior surface. Waltz and Andreesen developed a four-electrode disc array for high cervical placement, which was used in a number of cases.[140] The array was offered for modification of motor disorders under an investigational protocol for a period of time. Although the array coninues to be used in modification of pain, it has not proven to be sufficiently advantageous for alleviation of motor disorders to warrant the more extensive surgical procedure. The development of the functionally equivalent four-electrode catheter array (Table 1) which can be introduced percutaneously through a spinal tap needle seems to offer the advantages of the former disc array without the disadvantages of the more involved surgical procedure.

C. Stimulators

The selection of the appropriate internal stimulator system depends to a great extent on the needs of the patient. There are a variety of units on the market today, which are described in Table 2. The passive units are essentially similar electrically, but differ in size. These provide greater control of the stimulus and an easily replaceable power source (via the external battery) and are the likely first choice for patients who can tolerate and operate the external unit. The Neuromed unit (Figure 2) offers selection of the site(s) and polarity of stimulation which is useful in many situations. Totally implantable units developed thus far do not provide adequate flexibility in adjusting stimulation parameters, but are improving and may be useful in situations where external hardware is undesirable.

1. RF Coupled

a. Passive

The majority of patients in the past have been fitted with totally passive internal systems

Table 2
MANUFACTURERS OF SCS STIMULATORS

Manufacturer	Location	Stimulation equipment
Avery Laboratories	Farmingdale, N.Y.	RF-powered, capacitively coupled, monophasic, alternating phases
Medtronic	Minneapolis, Minn.	RF-powered, capacitively coupled, monophasic, self-contained (see Table 3),
Neuromed	Ft. Lauderdale, Fla.	RF-powered, capacitively coupled, mono- or alternating phases, programmable (see Figure 2)
Cordis	Miami, Fla.	Self-contained (see Table 3)

which derive their power and control through inductive coupling of radio frequency energy from an external source, by means of an air core transformer, loosely coupled across the skin. The major advantage of this device is its simplicity and hence its small implanted volume and inherent reliability. One of the disadvantages of this unit is the requirement for an externally worn power and control unit. The implanted units contain a tuned circuit to receive the energy, a diode to rectify it, and capacitors to smooth it and to provide the required no-net-DC current path. Figure 3 shows two such typical units, from Medtronic and Avery Laboratories. These units are neither constant current nor constant voltage, but this is of little consequence, since the load impedance presented by the body, which is in the range of 500 to 1500 Ω, is relatively constant in any given patient. However, since the strength of the stimulation depends on the transmitting-receiving antenna separation, care is required in maintaining proper placement of the antenna with such system. All active circuits and power sources are worn externally with these, providing adequate control of stimulation parameters with the exception of the polarity and site of stimulation.

More complex RF units have been constructed which produce pulses of alternating polarity[168] in order to minimize the risks inherent in the passage of current through body tissue. However, since the most effective depolarization of neural structures occurs at the cathode, use of the more complex alternating stimulators leads to less efficient stimulation, in that the optimally placed electrode (usually the proximal one) would provide stimulation with the correct polarity only 50% of the time. The simpler systems which employ capacitive coupling to minimize polarization effects have been found to be adequate. In the SCS literature, there are no reports of tissue damage in humans attributable to polarization effects following therapeutic levels of stimulation.

Perhaps a more serious drawback of the simpler systems (indeed, of essentially all the existing commercial systems) is the lack of information regarding the amount of current delivered to the SC. The open-loop nature of the devices and the dependence of the output amplitude on the antenna coupling make it impossible to calibrate the amount of current delivered to the electrode, let alone to measure the fraction of current passing through the cord itself.

b. Active

More recently, an active system which derives its operating power from the stimulation signal itself has been developed. This system allows the selection of any combination of up to four active electrodes, including selection of polarity, in addition to the usual control of stimulus amplitude, pulse width, frequency, and on-off control (Figure 3). This circuit requires the stimulation to be on for a certain minimum number of pulses and at a certain

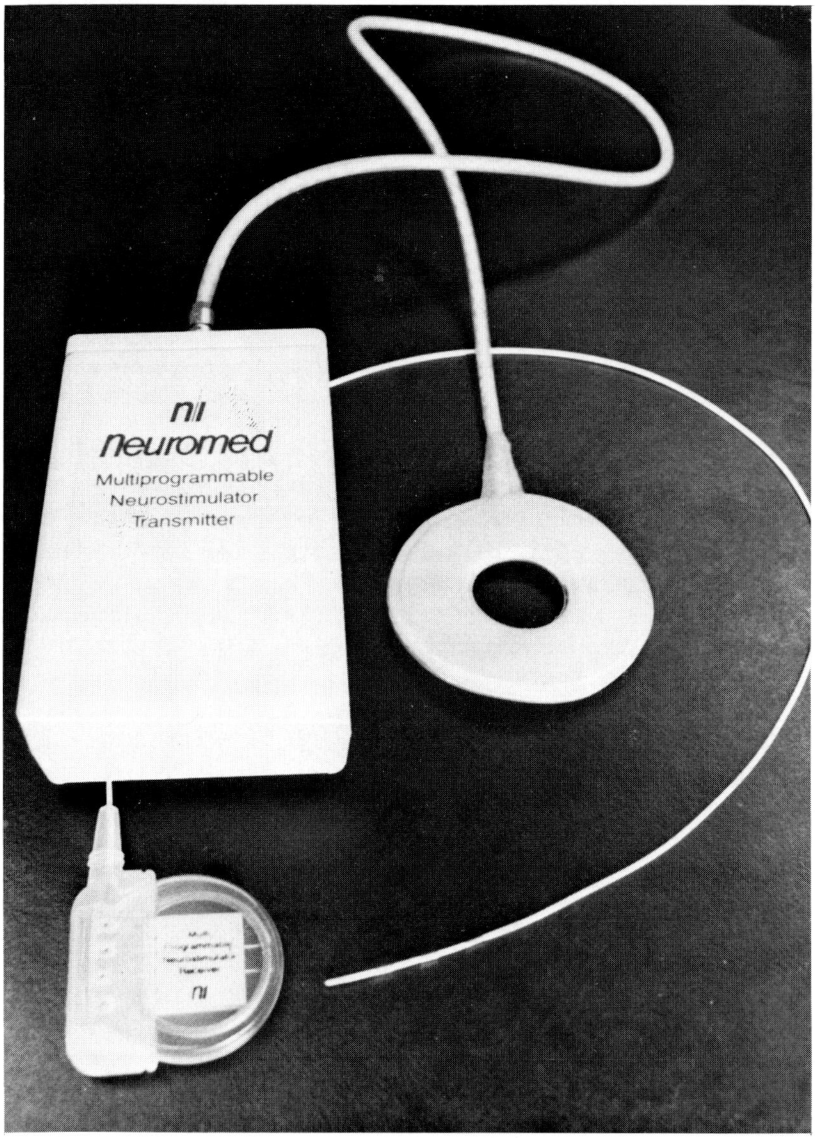

FIGURE 2. The multiprogrammable neurostimulator developed and produced by Neuromed represents a step forward in the evolution of SCS systems. It employs a four-electrode array on a single catheter connected to an active receiver capable of activating the four poles in any combination of polarities, in monophasic or alternating phase pulse trains over a range of frequencies from 25 to 1500 Hz. This allows simple external alteration of the stimulus distribution at any time to compensate for technical problems related to electrode slippage or to changed physiological requirements.

threshold level in order to ''program'' the internal circuitry to perform the desired functions. The control pattern is continuously retransmitted while the unit is operational to maintain control integrity. This unit has all of the advantages of the passive unit, except for the simplicity, and shares some of the disadvantages, particularly those regarding the antenna placement over the implanted receiver. Because of the necessity to power the internal circuitry, the unit cannot be operated at frequencies much below 10 Hz, unlike the passive systems. However, such low frequencies are not customarily employed in therapeutic stimulation.

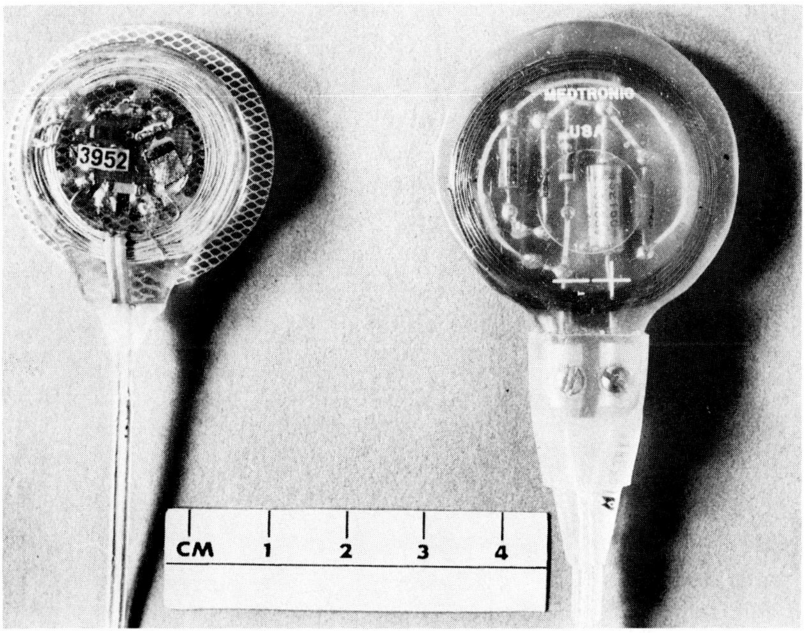

FIGURE 3. Passive, RF-powered SCS stimulators. In this figure are shown Avery (on the left) and Medtronic (on the right) implant units. Note the difference in thickness, as evidenced by the shadows.

Table 3
DESIGN PARAMETERS FOR SELF-CONTAINED IMPLANTABLE STIMULATORS

Parameter model, Manufacturer	Cordis 904	Medtronic
Frequency range	1—100 Hz	2—130 Hz
Steps	24	14
Stimulus intensity		
Constant current	0.8—8.5 mA	—
Steps	15	
Constant voltage	—	0—10 V
Increment	—	0.25 V
Pulse width	75—315 μsec	60—450 μsec
Steps	8	10
Control	Magnetic	Magnetic
Patient control	On/off	On/off
Projected life		
30 Hz, 0.2 msec, 5.5 mA	3.25 years	
30 Hz, 0.2 msec, 5 V		3 years
Electrode arrangement	Mono- or bifocal	Mono- or bifocal

2. *Self-Contained Implantable Stimulators*

Totally self-contained units based on pacemaker technology have been developed, initially by Cordis Laboratories and later by Medtronic, Inc. Difficulties to be overcome in such designs include the provision of adequate power and of adequate control over stimulation parameters. The power requirements for SCS are much greater than for cardiac pacing, due to the much higher duty cycle in SCS (as reflected in the relative frequency of stimulation — approximately 1 Hz in cardiac pacing and up to 100 Hz and above in SCS). As a measure of how well these problems have been overcome, Table 3 shows some representative pa-

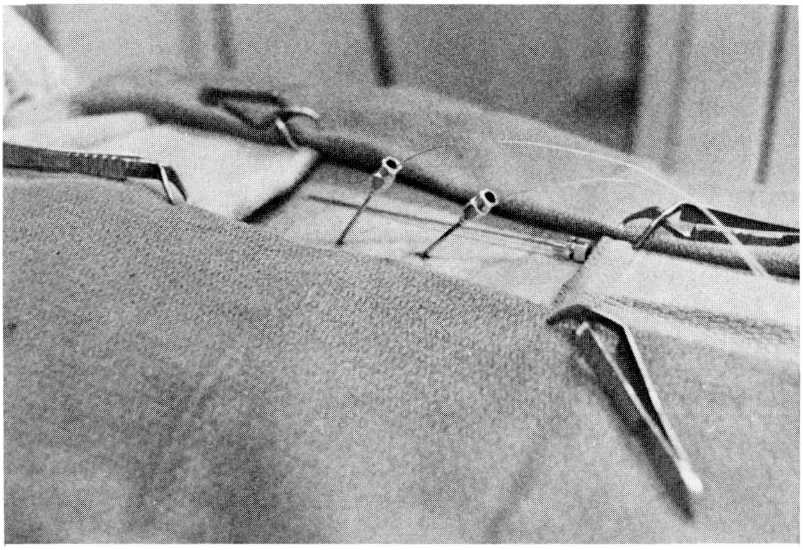

FIGURE 4. Percutaneous technique for placement of epidural SCS electrodes. Modified spinal Tuohy needles are inserted between vertebrae (in this case in the midthoracic region) to allow passage of the electrode into the epidural space.

rameters for these two systems. As a practical matter, the use of long pulse widths, high frequencies, and high current levels are contraindications for a totally implanted stimulator, in order to ensure reasonable life expectancy of the power source, regardless of the manufacturer. The second, and more resistant, problem is that of providing adequate control, particularly of stimulus intensity. In order to use these units, the responsible physician must have the appropriate external system able to control the implanted unit. Since the output levels have been discretized and since the required resolution in amplitude control is quite high, the basic control itself is poor. Many patients dynamically adjust the intensity of stimulation through the day, depending on their position (recumbent, sitting, and standing) and activity (sleeping, walking, and transferring from bed to wheelchair). At present, there are no convenient means for patients to intermittently adjust the stimulus intensity of totally implanted units.

D. Techniques for Placing Electrodes

1. Percutaneous Application of Epidural Electrodes

The most commonly used method for achieving SCS is through a percutaneous approach using a spinal tap needle, such as a Tuohy or Hustead needle, modified by enlarging the distal opening. After applying a local anesthetic, the needle with stylet in place is inserted through the skin in a cephalad direction into the interspinous space, through the ligamentum flavum, and into the epidural space. Placement of the needle in an appropriate orientation is verified using a fluoroscope. Penetration of the epidural space is identified by observing the flow of sterile saline placed in the needle into the epidural space. It is important to ensure that the needle is not subdural. When the needle is in the desired position, usually midline in the midthoracic region, the stylet is removed from the needle and the electrode is inserted through the needle into the epidural space (Figure 4). By careful selection of the insertion angle and of the rotation of the needle, the electrode lead can be directed rostrally along the posterior midline aspect of the SC. The space in the canal is created at least in part by the relative collapse of the veins in the canal when exposed to the higher atmospheric pressure, since the space is normally at a pressure slightly below atmospheric pressure. The formation of the space can be facilitated by the injection of saline. With careful technique, using fluoroscopic observation, the electrode can, in principle, be positioned wherever desired along the cord.

Normally, the leading electrode is placed in the midline of the lower cervical or upper thoracic portion of the spinal cord and the trailing electrode is placed one to three segments below it. In our work, the T1 vertebral location for the most rostral electrode has proven to be the most stable location from a technical point of view, to minimize electrode dislodgment and variation in stimulus intensity due to relative motion of the electrode and SC. It also appears to be a suitable position for affecting spasticity of the trunk and legs. If, on the other hand, the abnormal motor control is confined to the legs only, as in paraplegia, a lower placement, near the 12th thoracic vertebra, should be considered as an alternative stimulation location.

Once a satisfactory position is obtained as demonstrated by both anterior-posterior and lateral fluoroscopic views, the correct relationship to underlying neural structures should be verified by connecting the stimulator to the electrodes and depolarizing the underlying neural structures. Adequacy of placement of the electrodes with respect to viable neural structures can then be judged by report of the quality, intensity, and localization of the sensations of the patient and by observation of any muscle twitches evoked by low-frequency, relatively large amplitude stimuli.[40]

Upon successful placement of the electrodes, they are secured either by suturing to the skin if a trial period of stimulation is intended or by suturing to the underlying fascia through an incision around the base of the needle if a total implant is to be accomplished. Leads can then be connected to a suitable stimulation device as described below. For total implantation of the leads, either immediately or after a trial period, the leads must be passed through a s.c. tunnel to a distant point to preserve sterility. Lead extensions are available for temporary external stimulation.

2. *Laminectomy Application*

Due to the substantially increased surgical trauma involved, the laminectomy approach is less favored than is the percutaneous approach. However, there may be an advantage in the more invasive surgery, in that a more exact positioning of the electrode can be achieved and maintained through suturing to the dura, particularly if the spinal canal is extensively disrupted as a result of previous trauma. A modified laminectomy approach has been adopted for the insertion of an electrode array.[140] The procedure involves removing the appropriate spinous process entirely and 3 to 5 mm of the inferior lip of the same lamina. This creates an opening in the bony structure through which an incision is made to allow passage of the electrode array. When the array is in place, it is sutured to the dura. In this way, the position of the array relative to the SC may be better maintained. After fixation, the lead wires are brought out via a s.c. tunnel to an implanted receiver as described below.

3. *Trial Stimulation vs. Initial Implantation*

Disagreement exists regarding the desirability of an initial trial of stimulation with externalized electrodes. Cook et al. suggested such an approach since, even after 10 years of experience, it is still necessary to reject 40% of those tested with temporary stimulation.[29] Using the percutaneous approach, the placement of electrodes is relatively noninvasive and safe, with reasonable precautions under well-controlled clinical conditions.[29] On the other hand, others feel that such an evaluation is unfair, in that the patient is overly stressed during the evaluation and may not be able to give a reliable report.[30] A number of workers share this opinion and hence do not use the trial stimulation period as suggested by Cook. The development of multiple-lead electrode systems and implantable electronics make the total implantation option more desirable.[140] However, if extensive neurophysiological evaluation is desired, this can best be done with externalized leads.[37]

E. System Internalization

If the patient is selected for complete system internalization, the electrodes are put into position in the same manner as described above, but the surgery is extended to completely internalize the system in the same procedure. The lead wires coming from the electrode are passed subcutaneously to a suitable point for connection to the stimulus source. The stimulus source, either an RF receiver or battery-powered stimulator, is then placed in a s.c. pocket, usually in a location just below the rib cage, on the side. After testing is done to ensure system integrity, the lead wires are secured at the exit point from the intraspinous ligament, connections are carefully covered with insulation, and the incisions are closed.

F. Stimulation Parameters

Monophasic, capacitively coupled, pulses are commonly employed. A frequency of 33 Hz is typically used for screening, with 20 to 100 Hz used therapeutically. Several authors have reported that it is desirable to screen frequencies from 20 to 1400 Hz to find the optimal choice,[104,140] particularly in the treatment of spastic torticollis.[66-68] Pulse width is usually set at 100 to 200 μsec, unless increased to provide a stronger stimulus.

Stimulation intensities are normally adjusted to a level close to the sensory threshold. Measured pulse currents of 2 to 5 mA were commonly found for the sensory threshold at the time of initial testing with externalized systems. In some instances in which the electrodes were placed in the midthoracic region, and in spinal cord injury patients with impaired sensation, levels as high as 10 to 15 mA have occasionally been used. Such high levels of current seem to indicate less than optimal placement of electrodes with a consequent lower probability of success in modifying motor disorders.[85,114] As described above, the stimulus intensity could not be accurately assessed after total system internalization. Normally, the patient selects a level of stimulation which feels comfortable to him. This level is dependent upon the position of the electrode relative to various structures in the SC. When electrodes are placed in the cervical or high thoracic region, changes in position of the head and neck can change the effectiveness of the stimulation due to movement of the SC within the spinal canal. Much lower sensory thresholds were observed in those cases when the electrode was placed in close approximation to a nerve root or subdurally. In these latter instances, amplitudes of 1 mA or less were sufficient to produce excessive sensation and/or motor activity.

Stimulation is typically applied for 24 hr/day, with some variation in intensity according to the activities of the patient. Some find the stimulation distracting to the extent that it interferes with normal sleep, while others find it quite helpful to continue stimulation through the night to reduce the morning stiffness resulting from their spasticity. Many patients find it quite useful to control the intensity of stimulation in order to change the effect produced. This may be accomplished by turning down or switching off the stimulator. In other instances, patients have found that it may be useful to increase the stimulus level to a point where it actually increases the level of spasticity or rigidity for a period of time to accomplish a particular movement, for example.

G. Safety Considerations

1. Materials

As in any implant system, the electrodes and internal components must be constructed of materials which are nonreactive when implanted in the body. The system must be sterilized and free of toxic material such as plasticizers or other dissolved agents which might be absorbed from the system into body tissue. Furthermore, the electrodes themselves must approximate, to the maximum possible extent, the characteristics of an ideal, nonpolarizable electrode. Platinum seems to satisfy these requirements and is used in the majority of such systems at the present. Platinum-iridium has also been used to improve the mechanical characteristics of the electrodes, with no obvious compromise in other properties. The systems currently in use appear to be without major problems in the area of biomaterials compatibility, as is evidenced by at least one post-mortem examination study of four patients which showed no evidence of changes due to stimulation.[166] However, in a series of patients implanted with DCS systems for amelioration of pain, Pineda reported thickened arachnoid in 2 of 84 patients implanted over a 4-year period. These implants were done with the full laminectomy procedure, and implants were made subdurally.[174]

2. Surgical Risks

The surgical risks from SCS are low. If the percutaneous approach is used for placement of the electrodes, the risks are very low since the surgical trauma is lessened and the need for general anesthesia is obviated. MS patients may have some increased risk from the surgical procedure or possibly from the (general) anesthetic agent used,[145] due to the possibility of exacerbation of their disease process. Other studies have suggested the occurrence of such a relapse is not inevitable,[144] but that the occurrence of even a slight increase in temperature during or following surgery may increase the chances of a permanent degradation of the patient's condition.[153] The use of the catheter-tip electrodes and Tuohy needle placement in the epidural space requires only local anesthesia, as does the placement of the implanted stimulator itself. Procedures which utilize the externalized spinal leads have some increased risk of infection during the period that a transdermal pathway is maintained, but in practice, the risk can be adequately managed with reasonable precautions. There is a potential risk of epidural bleeding and a risk of spinal cord trauma from the spinal tap needle, the electrode, or pressure from a hematoma that could develop in response to invasion of the space with the electrode. However, no such cases have been reported, and the only bleeding observed appears to be without consequence.

3. Electrical Safety

The usual precautions with regard to shock hazards from leakage currents must be observed when working with SCS equipment. In general, we have not found it necessary to use these systems in patients with indwelling cardiac catheters. The permanently implanted systems have safety features inherent in their design: restricted output power, isolation from the external system by the RF coupling, and battery operation. Consideration of the maximum levels of current generated by these units rules out significant risks from direct heating effects of the current. There are no evident long-term risks from excessive depolarization of the cord itself, given that any induced pain or excessive muscular contraction would be evident and is easily controlled by reducing the stimulus intensity.

Consideration must be given to the electrochemical safety factors in SCS and the related tissue changes which might be so induced. The stimulating current is capacitively coupled to the electrodes, to prevent the polarization of electrodes and resultant build-up of toxic products which might result from DC stimulation. Using a 200-μsec, 10-mA pulse delivered through a typical electrode area of 0.08 cm^2, a charge density of approximately 25 $\mu C/cm^2$ is developed. This is well within the allowable limit according to the assumed damage mechanisms described by Brummer and Turner.[147] However, the resultant 2 μC per phase does exceed the 0.45 μC level reported to cause damage when in direct contact with brain tissue.[175] Epidural placement of electrodes would seem to obviate concern at this point, since the electrodes are not in direct contact with neural tissue. Bergveld[146] suggested that electrolysis would be avoided if constant current pulses are of sufficiently short duration to avoid the attainment of a constant voltage level. This may indicate that stimulus pulse durations should be limited to less than 1 msec.[85]

H. Technical Problems in the Application of SCS

1. Implant Failures

A number of problems have been encountered in SCS implant procedures, some of which have been due to insufficient aseptic precautions or improper surgical technique. A large number of earlier problems were caused by the use of poorly designed electrode lead systems (the Davis-Geck electrode, designed as a temporary cardiac pacing lead) which fractured after a short period of implantation in a high percentage of patients.[123] Some of the early designs of the disc electrode arrays suffered similar fatigue failure problems, which have been largely eliminated in the present versions. The helical coil electrodes, based on a proven cardiac pacing lead design, seem to have a good reliability record, with very few failures attributable to the electrode itself. Another problem area reported by several investigators has been the failure of external test leads connecting to implanted electrodes. This problem can be resolved by using heavier leads and by taking more precautions in handling the patients during the temporary stimulation phase. The largest number of problems with the electrodes has been related to dislodgment or movement of the electrode from its chosen stimulation site.[61] Instances have been reported in which the electrodes have worked their way entirely out of the spinal canal over a period of 6 months. In other cases, the electrodes have slipped together and shorted out. This problem has prompted at least two different engineering approaches to solutions, described above.

Davis and Gray reported a large number of failures of electronic components of SCS systems.[32] However, this seems to be the exception rather than the rule. Failure of implanted electronic components has been a minor problem in our series, occurring in only 1 out of 42 implants. Smith reported on a series of 208 DCS implants for relief of intractable pain in which the receiver failure rate was less than 0.5%.[184]

2. External System Failures

Failures of the external circuitry, on the other hand, are more common, due to the greater complexity of the circuitry and to the rough handling and wear and tear associated with daily use by patients with difficulties in performing skillful motor acts. The most common failure of equipment with any of the RF-coupled systems is the failure of the antenna, the link between the external box and the internal receiver. The antenna is typically subjected to rather severe stresses in the ordinary activities of daily living of these patients and most commonly suffered fatigue failures at the site of the connection to the stimulator. Smith reported a failure rate in excess of 90%/year for antennas and in excess of 12%/year for the transmitters.[184]

3. Technical Evaluation Procedures

The need for a technique for assessing the operational status of the implanted equipment independently of patient reports of sensation and independently of observed clinical benefits became evident, since sensation elicited by stimulation could originate from electrode surfaces located subcutaneously as well as epidurally and since clinical effects frequently exhibit rather long time constants. Radiologic observations of the electrode positions and lead integrity give some information, but do not define the functional status of the equipment. A method for mapping the electrical fields produced by the stimulator was developed which proved quite useful in localizing problems with the stimulation equipment.[123] In essence, the measured distribution of the stimulus artifact is compared with the expected distribution for the given electrode positions in order to determine if the system is functioning, if the polarity is as expected and if the system integrity has not been compromised through broken insulation or leads. When the above information is combined with patient reports and with physiologic evaluations (of evoked potentials and muscle twitches, described below), the technical adequacy of the stimulation can be established with a high degree of certainty.

V. EVALUATION OF THE EFFECTS OF SCS ON MOTOR PERFORMANCE

A. Clinical Evaluation

A large percentage of the reports of effectiveness of SCS in the literature is based on clinical observations of the patients made over a period of time ranging from 2 weeks to up to 5 years.[103,131] Some attempts have been made to use quantitative neurological assessment of various functions.[80,83,103] The use of film or videotape has been recommended to document the motor abilities of the patients before and after application of SCS.[90] However, the selection of appropriate evaluation methods according to the type of disorder being treated is still a major problem.

In addition to the modification of sensation (described in other chapters), three major types of effects have been reported following SCS: effects on motor control and spasticity, effects on bladder functioning, and effects on peripheral circulation. Virtually all authors have noted improvement in bladder function following SCS, but not all have found improvement in motor control. A number of other observations have been noted, from reduction in nystagmus to improvement in speech.[11] However, to date, selection of patients has been primarily on the basis of defects in motor control, bladder problems, or circulation.

It is well to note that DCS has been used for more than a decade and as such does not modify motor control in those patients with pain, but without motor disorders. It is only effective in replacing lost functions of motor control, functions related to sensory-motor integration, and the performance of motor acts. High levels of stimulation which directly activate motor cells, either orthodromically or antidromically, will produce motor output in both kinds of patients. These high levels are above those chosen for therapeutic stimulation, however, and such direct stimulation is not the desired result of therapeutic levels of stimulation.

1. Methods

The patient's condition is evaluated prior to, during, and if appropriate, following SCS. The methods used for the evaluation are selected according to the nature of the patient's disorder. To illustrate the selection of such tests, two patient groups, those with MS and those with SCI, are considered below. A factor which is frequently overlooked in such evaluation is the assurance that the stimulation is technically adequate and that the electrodes remain appropriately localized over the spinal column.

For MS patients, functional and neurological tests are performed, based on the following:

1. Cranial nerve functions with special stress on external ocular movement abnormalities
2. Active and passive range of movement, status of tendon reflexes, and tests for "pathological" reflexes of the upper and lower limbs
3. Strength, muscle tone, and degree of paralysis of individual muscle groups
4. Duration of clonus in muscles where it was elicitable
5. The patient's walking ability and posture, if possible
6. Alternating motion rate, fine finger movements, finger-to-nose or heel-to-knee when possible
7. Cutaneous and proprioceptive sensation[11]

For SCI patients, a different set of tests is used which is intended to assess the degree of completeness of the spinal cord lesion and the nature of the remaining pathways. Therefore, somatosensory evoked potentials from the legs as well as arms are evaluated. Finally, a comprehensive, polyelectromyographic (PEMG) examination of the function of the upper motor neuron is conducted.[156] Surface electrodes are placed over the muscle bellies of five muscles on each leg, and additional pairs are placed over abdominal and paraspinal muscles. With the patient in a comfortable, supine position, a standard protocol of maneuvers for activation of segmental reflexes and modification of them by descending influence from the brain is employed. Direct and indirect tests of volitional control, including the evaluation of activation of paralyzed muscles by reinforcement maneuvers, are carried out. In addition, tests of selected segmental reflexes which are dependent upon suprasegmental inputs are made, including the vibratory reflex and clonus.[154,155] We analyze the amount and distribution of EMG activity thus elicited and the consistency of the responses to three repeated trials.[180]

2. Motor Control and Spasticity

Motor control in man is the end result of the interaction of a number of different systems acting simultaneously and, in the healthy subject, results in an activation of muscle groups in an integrated, coordinated fashion. Disease or injury processes may result in a disintegration of these processes and thereby cause impairment of function. Although the functional outcome of the pathological process may be the same (e.g., unable to walk), the underlying physiological mechanisms responsible can be related to any combination of the following factors: loss of muscle power, contractures, loss of sensation, loss of upper motor neuron function controlling volitional motor activity, and loss of balance and postural control.

Improvement in motor control, then, may relate to any one or to a number of such mechanisms. One frequent and obvious change in motor function is the reduction of spasticity, occurring in MS and SCI patients following SCS.[35,103,131] Reduction of spasticity can in turn result in improved motor function, if the spasticity caused excessive coactivation of muscles and resulting fatigue. On the other hand, moderate amounts of spasticity may serve to stabilize otherwise uncontrolled joints during ambulation, the so-called ''spastic crutch''. Reduction of spasticity in these cases would actually lead to a deterioration of gait. Another change in motor function was increased endurance in ambulation and other activities of daily living.[7] Changes in strength, balance, walking, coordination, speech, swallowing, and eye movements were also reported in patients with MS or cerebellar degeneration.[31] A major problem in the evaluation of the effects of SCS relates to the fact that motor control is difficult to quantitate and is normally assessed only clinically. The clinical assessments have been challenged as being too subjective, and it has been suggested that objective, double-blind studies be done.[62] Investigators frequently find no significant changes in objective measures even though patients report subjective improvements in function. The lack of agreement between the patient's observation and objective findings is a consequence of the difficulty in designing valid tests of motor control. Objective measures which have been used include such things as tests of the H-reflex, the function recovery, and the ratio of H to M amplitudes. These tests, which focus on specific reflex pathways, are difficult to interpret in terms of changes in the patient's motor control. Other objective tests used, such as the somatosensory evoked potentials, have not yet been demonstrated to have specific relevance to motor control, although sensory function is undoubtedly an important aspect of motor control from a systems point of view.

In the PEMG examination which is an attempt to quantitate the degree of integration of the various systems involved in motor control, significantly different patterns of activity were observed before and after SCS. Voluntary activity tested after a period of SCS tended to be more controlled, with less inadvertent coactivation of inappropriate muscle groups and with better reciprocation in antagonistic groups. In the group of patients with SCI, passive maneuvers inducing long-lasting bilateral tonic bursts of activity before stimulation were much less effective in eliciting similar activity following SCS. Clonus was diminished or completely suppressed following SCS. The vibratory reflex response was less likely to spread to nonvibrated muscles following SCS in SCI patients.[40] Results of these tests did correlate with the clinical observation that SCS can affect muscle tone.[9]

3. Bladder Function

Bladder dysfunction is a common feature of severe upper motor neuron disorders in spinal cord injury and MS patients. Not only do such problems create serious social difficulties, they have also added substantially to the morbidity and mortality associated with upper motor neuron disorders. Essentially all investigators describing the effects of SCS have found that it did improve bladder control, especially with respect to decreasing the frequency and urgency of urination.[1,4,6,26,58,74,83,98,103] In a review of SCS literature, improvement in bladder control was found to be significantly improved in 40 to 50% of patients tested.[72] While it is possible to quantify the changes in bladder function with objective urological and urodynamic tests, this was not commonly done. One study of seven patients with stable bladder dysfunction due to spinal disorders showed objective clinical and urodynamic evidence of improvement of function.[98] This improvement reached a peak in 6 weeks to 2 months and persisted for at least 22 months. Another report noted that objective measurement of urine flow rate in patients before and after SCS showed a significant increase in MS patients and that the effect reached a peak 6 to 7 days following commencement of stimulation.[77] Urodynamic studies before and after SCS in seven MS patients with detrusor hyperreflexia showed an improvement in four.[83] Furthermore, several authors noted that the bladder status quickly returned to prestimulus levels if technical problems with the stimulation system interrupted the continuous application of SCS.[26,77] One report of objective assessment which failed to show improvement may be accounted for in part by the difficulty in achieving technically adequate stimulation (20 out of 23 patients had technical problems).[143] It should be noted that direct, subdural, intramedullary stimulation of the SC for control of bladder function has been used in several patients over a number of years.[173] Stimulation of the conus medullaris was used to induce bladder emptying in SCI patients, in contrast to SCS in MS patients which has been used to control frequency and urgency of urination. A modification of the percutaneous technique has been used to place one electrode epidurally and dorsal to the conus and to place another electrode intradurally and ventral to the conus region, to allow temporary application of SCS for bladder emptying during "spinal shock".[4]

4. Circulatory Changes

An experience common to most patients is sensation of warmth in their extremities within a short interval following the commencement of SCS. In anecdotal reports, it has been noted that skin conditions improve apparently as a result of improved circulation.[7,18,112] In 1 of our MS patients, skin ulcerations cleared up after about 6 months of SCS. Transcutaneous stimulation can have a similar effect on circulation.[148] In several patients, SCS was able to restore sufficient circulation to reverse signs that an extremity should be amputated.[4] They found that the stimulation need not be applied continuously for favorable changes in circulation to occur. A stimulation pattern of 1 hr "on" followed by 1 hr "off" in selected cases provided sufficient changes in circulation.[4] The change in circulation can be documented in a number of ways, but perhaps most clearly by thermography. Increases in peripheral circulation ranging from 75 to 150% were reported in another study,[58] in which improved functioning of the extremity showing the increased circulation was also seen. However, one objective study was unable to correlate the observed "spontaneous" changes in leg blood flow with onset or cessation of SCS.[77]

5. Other Changes

Other beneficial changes reflecting improved motor control of specific functions have been observed, particularly in MS patients. Ataxia is frequently found to diminish markedly upon application of SCS.[7,8,59,138] Improvements in balance, endurance, and walking were also commonly reported.[7,31] Speech improvements were also found in selected cases.[31] In one instance in our own series, a female patient with MS who had marked strabismus had a rapid (1- to 2-min) correction to nearly normal binocular vision. The rapid change caused her to complain at first that the SCS gave her "double vision", suggesting that her visual cortex did not accommodate quickly to the reacquisition of binocular vision. Several SCI patients have reported a return of sensation to dermatomes with marginal or poor sensibility adjacent to dermatomes with normal sensation.

6. Time-Course of Clinical Changes

It has been commonly observed that while some changes induced by SCS can be seen immediately, such as reduction of tremor and ataxia, others take time to develop, ranging from a few minutes, to 6 to 7 days (bladder), to as much as 3 months. It is difficult to explain such a wide range of time-courses with a single mechanism. It is more likely that a number of factors are operative. The longest time constant, 2 to 3 months, may be simply a reconditioning of patients who had been restricted in their activities through fatigue or spasticity and who, following a period of SCS, are able to resume some activities which previously had been inhibited by excessive spasticity. Such changes in life style were frequently noted in our series[11] and in others, who, for example, commented on the changes in required nursing care.[131] Various neurohumeral and endocrine changes have been proposed to explain the intermediate time results (1 to 7 days).[23,95]

Of further interest is the decay of the effect noted in a number of the patients when the stimulation was turned off or when technical failures in the stimulation system occurred. Patients reported that the sensation of stimulation did not completely disappear for periods ranging up to 4 to 8 hr. The effects on spasticity similarly persisted for 30 min to a few hours.

B. Biological Problems in SCS

1. Clinical Variability

Uncertainties in the precise clinical status of the patients have contributed to the difficulty in defining the clinical effects of SCS. In particular, the well-known variability in functional abilities from day-to-day and week-to-week associated with MS are very real and present complications in selection of patients (proper diagnosis) and in analysis of results. The possibility of further exacerbations of the disease also complicates the analysis of the long-term effects of SCS. In a similar fashion, the unknown extent of the pathomorphology in severe spinal cord lesions make it difficult to predict which patient will respond satisfactorily to SCS. While it is highly probable that residual pathways must be present in the damaged spinal cord for SCS to be successful, a precise definition of which pathways and to what degree they must be preserved is not yet available as a reliable guide to patient selection for SCS.

2. Placebo Effects and Unsuccessful Clinical Trials

An editorial in the *British Medical Journal* suggested that SCS used in MS may be simply a placebo and that "its efficacy or uselessness should be rigorously judged by objective and independent neurological observers".[62] A review of the available literature indicates that the changes in motor performance following SCS are modest as a rule, with exceptional cases of dramatic improvement. A significant number of patients continue to benefit from SCS after years of use. The inadequacy of measures available to document the changes in motor performance reported by the patients may add to the difficulty in assessment. For instance, in the report by Rosen and Barosum, they conclude that "SCS can no longer be recommended for MS patients" on the basis of their objective findings, but their objective findings show slight improvement, and five of nine subjectively reported at least minimal improvement.[108] Furthermore, their objective findings did not correlate with patient's subjective reports. Another indication that the reported results are not due to placebo can be seen in the fact that documented biochemical changes in response to stimulation have been observed,[23,87,95] along with other objectively measured functions such as the force of contraction.

The fact that the patient feels the stimulation makes a true double-blind test of efficacy very difficult. One paper attempting to overcome this difficulty made use of the fact that, since the stimulation is novel to the patient and of relatively unknown quality, placement of electrodes within the spinal canal at points other than over the posterior aspect of the cord would provide sensation, but a different physiological meaning.[117] Undoubtedly, in the multiple hundred cases reported in the literature, more than a few have had (unknowingly) just such misplacements. However, the difficulty was never discovered in most cases and has simply served as a confounding uncontrolled intervening variable. Placement at various levels up and down the cord, even though on the posterior surface, seems to demonstrate the same effects. Finally, technical failures which leave the patient with some sensation, such as a broken or dislodged electrode, have been frequently reported in the literature as having disrupted any beneficial effects of the stimulation.[61,77]

C. Neurophysiological Studies

The direct physiological effect of SCS can be easily observed in that it induces sensation, and sufficiently high intensities of stimulation have been found to result in gross muscle contractions. A number of investigators have systematically studied the effects of SCS on these and other aspects of physiological functioning. These effects fall into three broad categories:

1. Sensory effects
2. Motor effects
3. Conditioning effects

Indeed, even before the beginning of clinical applications of epidurally placed electrodes for SCS, electrodes were introduced into the epidural space through cathethers placed for spinal cord blocks, to record electrical events from the SC.[158,181-183] More recently, subarachnoid electrodes have been used to monitor spontaneous on-going activity of the SC.[159]

1. Sensory Evaluation

The patient's perception of the stimulation is normally employed in establishing therapeutic levels, as described above. In a systematic study of this perception in 32 patients at the time of initial placement of electrodes, the threshold for sensation of a continuous train of stimulation was found to range from 0.8 to 28 mA, with a mean of 6.7 mA and a SD of 5.2 mA.[39] In another study, it was found that patients who benefited from therapeutic SCS chose lower levels of stimulus current on average (8.75 mA) than did the group who did not benefit (20.7 mA),[86] which agrees with our observations.

Threshold-level stimulation is usually described as "tingling" or "buzzing" localized near the electrode site or at least in the stimulated dermatome. A displacement of the electrode away from the midline of the SC results in a corresponding shift in the perceived sensation. The effects of anode placement are more subtle, but can be measured with careful observation.[167] The quality of the sensation depends on the location of the electrodes to some extent, but much more on the frequency of stimulation. At rates of 1 to 10 Hz, the sensation is described as "heart-beat like". Position of the electrodes affects the distribution of the perceived sensation and the intensity much more than the quality. In several instances, the sensation was perceived (annoyingly) to wax and wane synchronous with the heart beat, until the electrode position was adjusted slightly. When the maximum tolerable stimulus levels were obtained in 19 patients, a value of 12.8 + 7.1 mA was found on initial placement, and a value of 10.8 + 4.4 mA was found upon reexamination. Both threshold and tolerance values were increased at 1 Hz compared to 30 Hz.[39] At intermediate frequences up to about 30 Hz, the sensation was "rough", changing to a "smoother" quality above 30 Hz. Little

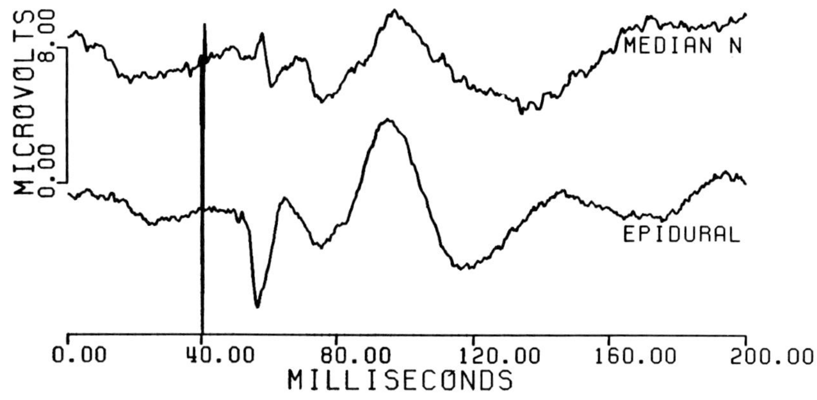

FIGURE 5. Evoked potentials from the scalp. A comparison of responses to peripheral nerve and spinal cord stimulation is shown in this figure, with the top trace showing a response to median nerve stimulation, and the bottom trace showing response to epidural stimulation. Both traces are the average of 128 responses recorded from C4-Fz EEG International Standard positions. Negative potentials are displayed upward in this figure and Figure 8. In each case, the stimulus occurred 40 msec after the beginning of the trace. The epidural electrodes were at the 12th thoracic and 1st lumbar vertebrae.

change in quality was noted above 50 Hz, but changes in frequency did change the threshold of sensation at these high frequencies. Although we have not done systematic studies with higher frequencies, patients reported that they did not continue to feel the high frequency stimulus after a few minutes. Sensory afferent signals evoked by noxious stimuli and recorded with microneurographic techniques have been compared with and without SCS in three patients.[161] In two of the three, SCS interfered with and partially blocked the noxious afferents. In the third, the electrodes were anterior.

Stimulation of the dorsal columns can evoke potentials recordable over the sensory cortex. A comparison of cortical responses evoked by peripheral nerve stimulation and by spinal cord stimulation is presented in Figure 5. The top trace, recorded from surface electrodes at C4-Fz positions (EEG standard nomenclature), was elicited by averaging 128 responses to stimulation of the median nerve at the wrist. The bottom trace, also from C4-Fz, was elicited by stimulation from electrodes in the epidural space at Cv3-Cv4 over the dorsal aspect of the spinal cord. In each case, the stimulus was a 0.5-msec pulse, adjusted to the twitch threshold for the thumb when stimulating the median nerve and to the sensory perception threshold when stimulating epidurally. When the electrodes are not situated over the dorsal aspect of the cord, but are instead ventral, the characteristic, short-latency, sharp, well-defined waveforms are not produced. Only low-amplitude, poorly defined waveforms are present, if at all in such cases.[50] By observing these characteristics, the location of the SCS electrodes relative to various structures of the spinal cord can be determined. We have found, for example, that assymetries in the distribution of cortical SEPs can be related to the location of the SCS electrodes laterally in the epidural space. In addition to the evaluation of the location of the electrodes, this technique can be used to demonstrate the viability of neural structures being stimulated by the SCS system.

2. Motor Studies

Electrical stimuli applied through posteriorly placed epidural electrodes give rise to twitch responses in the myotome corresponding to the epidural electrode location when the stimulus strength is adjusted to an adequate level[36] and less frequently in distant myotomes as well.[48] The twitch response latency was found to be approximately half of the stretch reflex latency in the corresponding muscle.[41] Figure 6 shows the EMG twitch response of an elbow

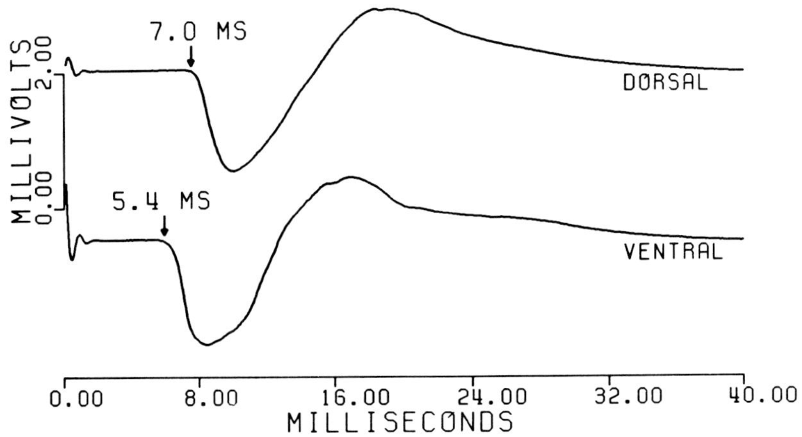

FIGURE 6. Muscle twitches elicited by SCS. With an epidural electrode over both dorsal and ventral aspects of the spinal cord at the third cervical vertebra, muscle twitch responses in the biceps brachii muscles to stimulation of the two aspects of the cord can be compared. The top trace shows the average of eight responses elicited by stimulation of the dorsal, and the bottom trace of the ventral sides of the cord.

flexor muscle to single epidural stimuli. The top trace shows the muscle response when stimulated from an electrode over the dorsal aspect of the spinal cord and the bottom trace shows the response of the same muscle when stimulated from an electrode over the ventral aspect at the same level. In both cases, the anode was the same skin surface electrode. Note that stimulation of the ventral aspect resulted in a shorter response latency. When repetitive stimuli were applied through posteriorly placed electrodes, the muscle continued to respond with a synchronized burst up to a stimulus frequency of approximately 10 Hz, after which a poorly synchronized response was found. On the other hand, repetitive stimuli of 50 to 100 Hz still elicited synchronous, 1-to-1 EMG responses when applied to the ventral side of the spinal cord. When stimulating through the epidural electrodes, a comparison of the right and left side EMG thresholds and amplitudes will reveal whether the electrodes are symmetrically placed.[119] When stimulus intensities were adjusted to at least twice the threshold value for the segmental response, distant motor twitches were also elicited in 14 of 70 patients. In most cases the distant response was elicited only in the short toe flexor muscle and only occasionally were other muscles activated (Table 4).[48]

The use of SCS electrodes for direct, systematic measurement of spinal cord functioning is being exploited through the development of a new epidural electrode. This new system for epidural spinal cord evaluation (SESCE) was designed for temporary application only. It has multiple active electrode elements and a second lumen parallel to the electrode leads for the induction of chemical agents or removal of spinal fluid samples.[47] The system can be used to conduct examinations of sensory and motor systems described above, as well as to apply temporary therapeutic stimulation. This development of the capability for direct measurement of SC function offers hope of substantially improving diagnostic and functional evaluation of the SC and hence of improving the selection of therapeutic approaches to the restoration of function in the damaged cord.

3. Conditioning Studies

Because of the difficulty in making and interpreting clinical observations related to the effects of SCS on motor function, a number of groups have chosen to evaluate a variety of other physiological parameters such as changes which occur in evoked potentials, in the contingent negative variation potentials, in H-reflexes, and in tendon tap responses as a

Table 4
PATIENTS EXHIBITING DISTANT MOTOR RESPONSES TO SCS

Diagnosis	Electrode level	Muscles activated	Response latency (msec)
MS	T3 - T4	Right quadriceps	17.0
		Left quadriceps	17.0
		Right triceps surae	25.0
		Left triceps surae	26.0
SCI-C6	T2 - T4	Right short toe flexor	42.0
		Left short toe flexor	42.0
SCI-C5	T2 - T4	Right short toe flexor	42.0
		Left short toe flexor	40.5
HI	C2 - C3	Right short toe flexor	40.0
HI	C2 - C4	Left short toe flexor	45.0
SCI-C6	T3 - T4	Bulbocavernosus	47.5
SCI-C4	T3 - T4	Right short toe flexor	40.5
		Left short toe flexor	41.0
MS	C7 - T1	Right quadriceps	20.0
		Right triceps surae	28.0
SCI-C5	T2 - T3	Right triceps surae	39.0
SCI-C6	T3 - T4	Right short toe flexor	40.5
		Left short toe flexor	41.5
SFP	C7 - T2	Right tibialis anterior	27.5
		Left tibialis anterior	27.0
		Right triceps surae	28.0
		Right short toe flexor	40.0
		Left short toe flexor	41.5
		Right short toe extensor	42.0
		Left short toe extensor	44.0
SCI-C4	T1 - T2	Right short toe flexor	40.0
		Left short toe flexor	41.0
SCI-C3	T2 - T3	Right short toe flexor	38.8
		Left short toe flexor	40.5
SCI-C6	T2 - T3	Left short toe flexor	43.0

result either of immediate or long-term effects of SCS. Measurements made before and after at least 4 days of SCS suggested a distinct increase in amplitude in the earlier (spinal cord) component of the cervical somatosensory evoked potential (SEP) in seven of eight MS patients and an increase in later components in 4 of the same patients.[113,114] These changes, however, did not correlate with clinical changes. In this same study, evaluation of the contingent negative potentials failed to reveal any evidence of psychological conditioning which might account for other changes.[113] This same group found similar improvements in brain stem auditory evoked responses, but not in visually evoked potentials.[113] In another study, only one of eight MS patients tested showed a change in SEPs, in that the one had measurable SEPs following SCS which was absent before.[87] Another author reported no change in cerebral event related potentials with SCS, even in patients with good clinical effects.[2,3] Other studies found modest changes in the SEPs in two of four patients with good clinical results of SCS.[37,73]

H-reflex changes during stimulation of patients with upper motor neuron disorders have been examined by several authors.[65,78,110,129,134] In one study, the H-reflex excitability threshold and the ratio of H to M waves were not consistently changed in a group of four MS and one SCI patients, nor were the results of H1-H2 conditioning studies consistently

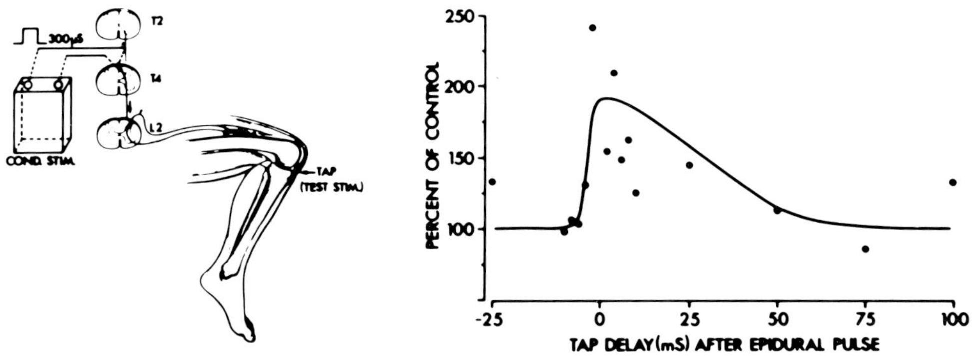

FIGURE 7. T-wave conditioning with epidural stimulation. Conditioning pulses were applied through epidural electrodes with the cathode at T2 and the anode at T4, at a 1-Hz stimulus rate, and were increased in amplitude until the conditioning effect was seen, but before the conditioning became generalized, i.e., below a level producing long-lasting facilitation. The procedure is shown schematically on the left. By varying the interval between the conditioning pulse and the tap, the response on the right was found in this SCI patient. Each point is the average of 20 responses and was compared to 20 control taps without the conditioning.

changed, except that the H2 response at intervals between 20 and 100 msec was closer to the control amplitude following SCS.[110] A second author reported on a similar study in an MS patient before and after SCS without the change described above, but with a normalization of the recovery curve which persisted over a 2-year period of stimulation.[65] In an earlier study, a third author showed the same changes occurring over a short period of time in one patient and a shift towards normal in the influence of cutaneous stimulation on the H reflex amplitude in a second patient.[78] Both the second and third authors reported that the H-reflex changes seemed to correlate with changes in the patient's clinical condition.

The immediate conditioning effects of epidural stimulation on segmental reflexes can be easily demonstrated. One example of such effects can be seen in Figure 7 in the modification of tendon tap responses with appropriately timed single pulses applied through the epidural electrodes.[37] Other changes can also be found, such as the facilitation of the tonic vibratory reflex response with continuous stimulation.[37,41] In addition, it was possible to show that peripheral inputs could modify motor responses to epidural stimulation. This was particularly true in the case of muscle twitches elicited in myotomes remote from the stimulated segment of the spinal cord. When conditioning stimuli in the form of cutaneous taps were applied at various intervals before the epidural stimulus, the largest amount of facilitation was present when the cutaneous stimulus proceeded the epidural stimulus by 40 msec. This effect was short lasting and disappeared after 70 msec. A unilateral facilitatory effect was also observed when vibration was applied to the plantar surface. When we studied the effect of reinforcement maneuvers in SCI patients with no volitional control, we found that the patients could facilitate and suppress the twitches with voluntary efforts to flex and extend the toes, even if they had little or no voluntary movement otherwise.

Direct measurement of muscle tone and spasticity can also play an important role in the evaluation of SCS effectiveness, using, for example, the isokinetic dynamometry approach to this measurement.[6,93] In one study, two spastic patients improved in voluntary strength and had diminished resistance to passive movements as well.[93] This method is relatively quick and simple to apply and offers hope of objective documentation of the patient's muscle tone and power at the time of testing.

4. Epidural Recording

Shimoji et al. studied spinal cord potentials evoked by segmental nerve stimulation as well as by ascending and descending volleys along the spinal cord in man. He recorded from the posterior epidural space at the cervical and lumbar enlargement levels and reported that it was easy to use such procedures in both patients and healthy volunteers.[181] Similar records have been made using electrodes placed for temporary SCS studies, with externalized distal ends.[178, 180] Epidural recordings have a much higher frequency content than do surface recordings, due to their proximity to the generators of the potentials. If recordings are made from the anterior portion of the cord, the recorded responses are much smaller.[178] At least three distinct components can be seen in posterior epidural electrodes in the cervical region in response to median nerve stimulation. They are

1. A large, widespread negative potential occurring at a latency of 13 msec
2. Smaller, very fast, traveling waves with progressively longer latencies proceeding rostrally, ranging in latency from about 10 to 14 msec
3. A slower (longer latency and longer duration) positive potential with an approximate latency of 16 to 22 msec

The large negative and slow positive waves are related to synaptic depolarization occurring as the nerve fibers enter the cord, while the faster waves appear to mark the passage of the impulse directly up the cord without synaptic intervention. This observation is borne out by studies of the refractory characteristics of the various components.

Potentials recorded from electrodes in the neck region are markedly different in response to stimulation of the arms as compared to stimulation of the legs (Figure 8). Stimulation of the median nerve gave rise to the complex potential described above. In contrast, bilateral tibial nerve stimulation resulted only in a much smaller conduction wave. Attempts to record conduction potentials in patients with spinal cord injury resulted in reduced amplitude or absence of certain components of the expected waveform.

VI. MODELING STUDIES

Since the underlying neurophysiological mechanisms responsible for beneficial effect of SCS are not known, it is not clear which, if any, animal models are relevant to the understanding of SCS. However, a biophysical study of the nature and distribution of electric fields produced within the spinal cord and of the effects of these fields on neuronal structures in the cord can contribute to the understanding of SCS.

Attempts have been made to measure the current distribution within the spinal cord produced by various electrode arrays and to infer from these results their relative effectiveness in stimulating the intended neural structures.[185,186] While such studies are tedious and fraught with experimental difficulties, they are essential in providing a basis of comparison of different electrode arrays. Building on this experimental data base and on other available data which describe the electrical conductivity of various hard and soft tissues[162] and using the finite element analysis techniques developed during the last few years,[189] several authors have studied the pattern of distribution of current and voltage fields when current is passed between electrodes in the vicinity of the human spinal cord.[149,152,176] These studies have taken advantage of the fact that, with a few important exceptions, the body can be treated as a homogeneous conducting medium and that phase shifts arising from transient phenomena in electrical fields are insignificantly small, allowing the problem to be treated as a steady-state problem. Coburn presented a two-dimensional model in 1980,[149] in which he described the voltage fields to be expected from midline and from dorsolateral electrode placements. He postulated that the major inhomogeneities of importance in modifying the current density are the CSF-fat interfaces and that the irregular bone geometry, particularly variations in the vertical axis, may not be important. In a subsequent presentation,[151] he described the results of experimental measurements made in patients which supported the simplifying

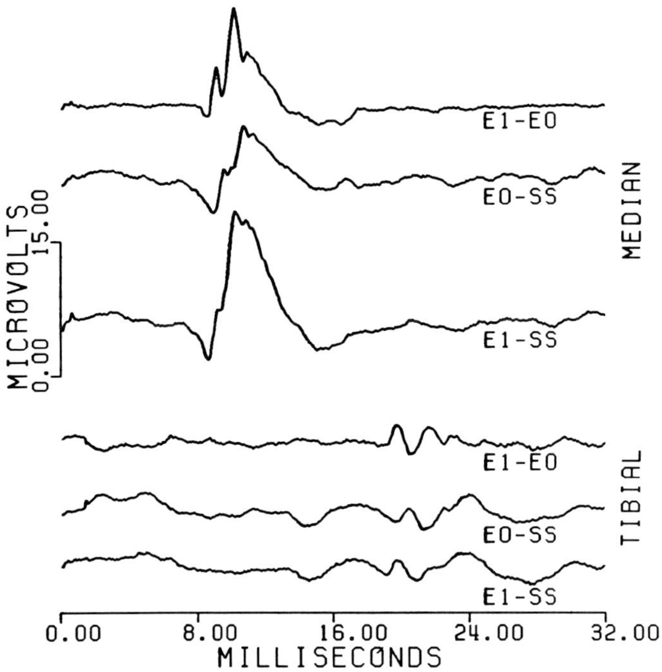

FIGURE 8. Epidurally recorded evoked potentials. Epidural electrodes at the fourth and fifth cervical vertebrae in a cerebral palsy patient were used to record the responses to median nerve stimulation in the top traces and to tibial nerve stimulation in the bottom traces. The three traces in each set are E1-E0 — bifocal pairing of the two epidural electrodes, E0-SS — monofocal recording of the top (C4) epidural electrode to an external references (SS) electrode at the suprasternal notch, and E1-E0 — monofocal pairing of the lower (C5) electrode to SS. Note that the median response is larger due to the activation of segmental structures, seen best in the two monofocal leads, along with the conduction wave, seen best in the bifocal pairing. In contrast, the tibial stimulation response was seen only as a conduction wave, with no large amplitude, slower segmental response, as could be expected by the entry of the stimulated nerve roots into the spinal cord at a distance from the recording electrodes.

assumption of vertical homogeneity in the patterns of current flow and hence of a prismatic model for the spread of current. In this latter report, he also described the results of a three-dimensional analysis in which the calculated impedance was within a factor of two of that typically reported from measurements made in patients. He reported the results of application of theoretical distributions of current derived from three-dimensional model studies to a long myelinated fiber model proposed by McNeal.[171] These calculations, for hypothesized dorsal column fibers 5 and 10 μm in diameter, predicted strength-duration curves for firing thresholds which bracketed strength-duration curves measured in patients. The clinical data were obtained by setting various pulse widths and asking the patient to adjust amplitude to sensory threshold.[151] Similar studies of the shape of the field distributions during monofocal and bifocal stimulation suggested that, for depolarization of long ascending fibers, multiple electrodes near the spinal cord behave essentially independently in effecting excitation, as long as they are separated by more than 1 cm.[151]

Coburn also presented examples of the computed electric field strengths at various points in the model spinal cord and showed that it was unlikely for cells and dendritic tree processes to be excited by such stimulation.[149-151] On the other hand, long fiber structures, such as nerve roots and long ascending or descending fibers in appropriate regions can be excited, particularly if they make a sharp bend in the region of higher field strength.[149-151] These findings support the general assumption that the primary target of stimulation is the ascending dorsal column fibers, with other effects due to nerve root depolarization.

VII. HYPOTHESIZED MECHANISMS

Patients undergoing therapeutic SCS as described in this chapter have a common problem in the loss of upper motor neuron functioning. They may have other neurological deficits as well, which they may share with the whole group, or which may be unique. The origins of the upper motor neuron dysfunction also vary. Since SCS is physiologically active, i.e., it can depolarize neural structures in the SC, the question is what aspect (or aspects) of the stimulation is important in the therapeutic effects seen in patients with upper motor neuron dysfunctions? The fact that stimulation of the spinal cord could reduce hypertonicity in decerebrate cats was described in 1902 by Frolich and Sherrington.[161] It is not clear how SCS works in man, however.

A wide variety of possible mechanisms involved in generating the therapeutic effects of SCS have been suggested; one reference lists 36 different proposed mechanisms.[72] With few exceptions, most authors rule out psychogenic effects due to the nature of the observed effects and to physiological observations. Biochemical changes have been suggested, but no convincing evidence has been presented. Activation of the autonomic nervous system has been proposed, and indeed, many changes described above may be due to such activation. Finally, depolarization of orthodromic and antidromic pathways in the somatic nervous system have been proposed.[132]

It can be hypothesized that the effect of SCS is predominantly indirect, through activation of the posterior columns and, secondarily, brain stem structures. As described by Wall and Noordenbos[187,188] the function of the dorsal columns is not well understood, but appears to be related to the ability to simultaneously analyze spatial and temporal aspects of information concerned with exploration and identification of objects external to the body and body surface. Activation of the dorsal columns and related brain stem structures may modify descending motor volleys and hence influence segmental reflexes through mechanisms described by Lundberg.[170] This hypothesis can account for the large degree of plasticity observed in the effects of SCS on muscle tone regulation, suppression of cutaneo-muscular reflexes, and improvement of coordination of movement in a large variety of motor disorders.

Two implications of this hypothesis of indirect action via the dorsal columns have been borne out by our observations of patients being treated with SCS. First the indirect influence of the SCS is widely distributed through the CNS. This generalization is substantiated by observations of the effects in MS patients, who exhibit different kinds and degrees of motor disorders, and different degrees of recovery. Second, the stimulation does not generate any direct motor responses, but rather modifies ongoing sensory/motor integration. This indirect nature is indicated in the modification of spasticity in spinal cord injury patients. An important fact in trying to understand the underlying mechanisms of SCS is that no completely absent functions can be generated, although hidden functions may become apparent following facilitation of the function or suppression of spasticity.

The proposed indirect path for SCS assumes that the integrity of the brain and the posterior, lateral, and anterior spinal columns is preserved, at least to some degree. Even though the functions mediated via the pathways are impaired due to the demyelination process in MS, it is evident that, in many instances, there are sufficient numbers of preserved pathways available to transmit information generated by the stimulation. This is particularly true in the moderately advanced cases considered to be the best candidates for epidural stimulation. While the damage to nerve fibers in the various pathways may be more severe in the degenerative disorders, some proportion of fibers will be preserved, providing paths for the modification of motor control.

In paralyzed spinal cord injury patients, it might be assumed that there are no ascending and descending paths remaining to convey the indirect stimulation effects. However, even in spinal cord injury patients with no clinical evidence of functioning spinal pathways, both neurophysiological evidence of brain influence on segmental reflexes[156] and neuropathological evidence of the existence of those pathways[164] can frequently be found. The preserved pathways conveying this influence may also convey the indirect effects of the SCS. One of the criteria we have used in the selection of SCI patients for SCS is evidence of partially preserved ascending and descending pathways. Under the assumptions of this model, SCS could not work in the same way in patients without these pathways.

It may be asked what effects could be expected if the posterior tracts are damaged or absent due to the nature of the lesion. When the posterior column cannot transmit ascending volleys to reach brain structures which are relevant in the modification of motor control, then another, simpler system could potentially play the same role, namely, the propriospinal interneuron system.[160] This system is much simpler and does not ascend further than the medulla oblongata. It extends diffusely through the spinal cord and impulses transmitted through the propriospinal system can modify the activity of motorneuron pools within the spinal cord. However, in the two SCI patients with absence of sensation and absence of cortical evoked potentials in response to stimulation of nerves of the legs, SCS produced no demonstrable changes in their abnormal motor control. It has been frequently noted that the diagnosis of Strumpfel's familial spastic paraplegia (FSP) is a good indication to use SCS to improve motor control.[7, 11,59,61,131] The recent observation that the dorsal columns are also affected in FSP[157] suggests why SCS is useful and lends further support to the idea that depolarization of dorsal column structures is important.

Finally, the clinical findings that the effects of SCS take days to be fully developed and that the effects may persist for hours to days after the cessation of stimulation suggest that the effects are not due to stimulation of just one structure. This is further supported by the observation that the effects are diffuse. Spasticity of both flexor and extensor muscle groups can be reduced, and the functioning of spastic bladders can be improved as well. All these observations point to the development of an altered sensory/motor integration of the brain and spinal cord structures.

ACKNOWLEDGMENTS

I would like to express my gratitude for the major contributions of my co-workers at TIRR for their assistance in this project, which made it possible for me to compile this summary. In particular, I am indebted to Dr. Milan Dimitrijevic, for sharing many ideas and for his most constructive critique of this manuscript. Dr. Faganel has provided invaluable insight in discussions of the clinical and neurophysiological aspects of this chapter. In addition, to Drs. Sharkey, Campos, Meta Dimitrijevic, Dahm, and the technical staff who supported them in the SCS efforts goes the credit for the SCS program for abnormal motor control at TIRR. Finally I am grateful to the Vivian L. Smith Foundation for Restorative Neurology for major financial support and to the Rehabilitation Research and Training Center No. 4 (Rehabilitation Services Administration Grant 16-P-56813-6) and Rehabilitation Services Administration Grant 13-P-59275-6 for their support.

REFERENCES*

1. **Abbate, A. D., Cook, A. W., and Atallah, M.,** Effect of electrical stimulation of the thoracic spinal cord on the function of the bladder in multiple sclerosis, *J. Urol.*, 117, 285, 1977.
2. **Abraham, P., Docherty, T., and Spencer, S.,** Cerebral event related potentials and spinal cord stimulation, *Electroencephologr. Clin. Neurophysiol.*, 44, 132, 1978.
3. **Abraham, P., Docherty, T., Spencer, S., Cook, A. W., Oygar, A., Illis, L. S., and Sedgwick, E. M.,** Spinal cord stimulation and event related potentials, in Multidisciplinary Perspectives in Event Related Brain Potential Research, Otto, D. A., Ed., EPA 600/9-77-043, Environmental Protection Agency, U.S. Government Printing Office, Washington, D.C., 1977.
4. **Augustensen, L. E., Carlsson, C. A., and Fall, M.,** Autonomic effects of electrostimulation, *Appl. Neurophysiol.*, 45, 185, 1982.
5. **Bates, J. A. V.,** Therapeutic electrical stimulation. The transistorized placebo?, *Contemp. Clin. Neurophysiol.*, 34(Suppl.), 329, 1978.
6. **Berg, V., Bergman, S., Hovdal, Hunstad, N., Johnsen, H. J., Levin, L., and Sjaastad, O.,** The value of dorsal column stimulation in multiple sclerosis, *Scand. J. Rehab. Med.*, 14, 183, 1982.
7. **Campos, R. J., Dimitrijevic, M. M., and Sharkey, P. C.,** Clinical evaluation of the effects of spinal cord stimulation on motor performance in patients with upper motor neurone lesions, in *Advances in External Control of Human Extremities*, Yugoslav Committee for Electronics and Automation, Belgrade, 1978, 559.
8. **Campos, R. J., Dimitrijevic, M. M., Dimitrijevic, M. R., and Sharkey, P. C.,** Effects of continuous spinal cord stimulation of the thoracic posterior portion of the spinal cord in patients with severe ataxia, *Acta Neurol. Scand.*, 60(Suppl.), 73, 223, 1979.
9. **Campos, R. J., Dimitrijevic, M. M., Dimitrijevic, M. R., and Sharkey, P. C.,** Suppression of decerebrate rigidity in patients with brain stem lesions by continuous stimulation, *Acta Neurol. Scand.*, 60 (Suppl.), 73, 224, 1979.
10. **Campos, R. J., Dimitrijevic, M. M., and Sharkey, P. C.,** Clinical evaluation of the effects of spinal cord stimulation on motor performance in patients with motor neurone lesions, in *Bulletin of the Los Angeles Neurological Societies*, McIntyre, H., Ed., Proc. Federation of Western Societies of Neurological Sciences, Scottsdale, Ariz., 1979.
11. **Campos, R. J., Dimitrijevic, M. M., Faganel, J., and Sharkey, P. C.,** Clinical evaluation of the effects of spinal cord stimulation on motor performance in patients with upper motor neurone lesions, in Proc. 3rd Int. Meeting on Spinal Cord Stimulation, Houston, 1980, *Appl. Neurophysiol.*, 44, 141, 1981.
12. **Cook, A. W. and Weinstein, S. P.,** Chronic dorsal column stimulation in multiple sclerosis, *N.Y. State J. Med.*, 73, 2868, 1973.
13. **Cook, A. W.,** Regeneration in the central nervous system (letter), *Lancet*, i, 1442, 1973.
14. **Cook, A. W.,** Electrical stimulation of the spinal cord (letter), *Lancet*, ii, 869, 1973.
15. **Cook, A. W.,** Stimulation of the spinal cord in motor neurone disease, *Lancet*, ii, 230, 1974.
16. **Cook, A. W.,** Electrical stimulation of the spinal cord, *Br. J. Pharmacol.*, 50, 451, 1974.
17. **Cook, A. W.,** Electrical stimulation in multiple sclerosis, *Hosp. Pract.*, 11(4), 51, 1976.
18. **Cook, A. W., Oygar, A., Baggenstos, P., Pacheco, S., and Kleriga, E.,** Vascular diseases of extremities, electrical stimulation of spinal cord and posterior roots, *N.Y. State J. Med.*, 76, 366, 1976.
19. **Cook, A. W.,** Percutaneous trial for implantable stimulating devices, *J. Neurosurg.*, 44, 650, 1976.
20. **Cook, A. W.,** Epidural spinal cord RF stimulation in M.S., I.E.R.E. Golden Jubilee, Conf. Applications of Electricity in Medicine, April, 1976.
21. **Cook, A. W., Tahmoureses, A., Oygar, A., Kleriga, E., Pacheco, S., Baggenstos, P., and Nidzgorski, F.,** Epidural electrical stimulation of spinal cord for intractable pain and other abnormal conditions, *Acupunct. Electr-Ther. Res. Int. J.*, 2, 259, 1977.
22. **Cook, A. W.,** Spinal cord stimulation in multiple sclerosis, *Acupunct. Electr-Ther. Res. Int. J.*, 3, 265, 1978.
23. **Cook, A. W., Nidzgorski, F., Hurwic, M., Winsberg, B., and Klutch, A.,** Monamine accumulation among sclerotic patients receiving dorsal column stimulation, *Lancet*, i, 46, 1978.
24. **Cook, A. W., Nidzgorski, F., and Oygar, A.,** Chemical modulation of motor control with and without electrical stimulation of the spinal cord in multiple sclerosis, in *Advances in External Control of Human Extremities*, Yugoslav Committee for Electronics and Automation, Belgrade, 1978, 621.
25. **Cook, A. W., Taylor, K., and Nidzgorski, F.,** Epidural cathodal stimulation of the spinal cord, temporary and permanent, in *Advances in External Control of Human Extremities*, Yugoslav Committee for Electronics and Automation, Belgrade, 1978, 711.
26. **Cook, A. W., Abbate, A., Atallah, M., Pacheco, S., Kleriga, E., Saada, S., and Nidzgorski, F.,** Neurogenic bladder: reversal by stimulation of the thoracic spinal cord, *N.Y. State J. Med.*, 79, 255, 1979.

* References 144 to 189 are included for this text only and are not directly related to SCS for motor disorders in humans.

27. **Cook, A. W., Taylor, J. K., and Nidzgorski, F.,** Functional stimulation of the spinal cord in multiple sclerosis, *J. Med. Eng. Tech.*, 3, 18, 1979.
28. **Cook, A. W., Taylor, J. K., and Nidzgorski, F.,** Epidural cathodal stimulation of the spinal cord, temporary and permanent: a short report, *Int. Rehab. Med.*, 2, 68, 1980.
29. **Cook, A. W., Taylor, J. K., and Nidzgorski, F.,** Results of spinal cord stimulation in multiple sclerosis, in Proc. 3rd Int. Meeting on Spinal Cord Stimulation, Houston, 1980, *Appl. Neurophysiol.*, 44, 55, 1981.
30. **Davis, R., Flitter, M. A., Bolton, D., and Komrad, M.,** Clinical efficacy and safety of chronic spinal stimulation used in multiple sclerosis and other demyelinating diseases, in *Advances in External Control of Human Extremities,* Yugoslav Committee for Electronics and Automation, Belgrade, 1978, 557.
31. **Davis, R., Gray, E., and Kudzma, J.,** Beneficial augmentation following dorsal column stimulation in some neurological diseases, in Proc. 3rd Int. Meeting on Spinal Cord Stimulation, Houston, 1980, *Appl. Neurophysiol.*, 44, 37, 1981.
32. **Davis, R. and Gray, E.,** Technical factors important to dorsal column stimulation, in Proc. 3rd Int. Meeting on Spinal Cord Stimulation, Houston, 1980, *Appl. Neurophysiol.*, 44, 160, 1981.
33. **DeJong, R. N. and Sugar, O.,** *The Year Book of Neurology and Neurosurgery,* Year Book Medical Publ., London, 1981, 168.
34. **Delwaide, P. J., Toulouse, P., and Crenna, P.,** Hypothetical role of long-loop reflex pathways, in Proc. 3rd Int. Meeting on Spinal Cord Stimulation, Houston, 1980, *Appl. Neurophysiol.*, 44, 171, 1981.
35. **Dimitrijevic, M. R., Sherwood, A. M., and Faganel, J.,** Mechanisms of motor control augmentation using continuous epidural spinal cord stimulation, in *Advances in External Control of Human Extremities,* Yugoslav Committee for Electronics and Automation, Belgrade, 1978, 657.
36. **Dimitrijevic, M. R., Janko, M., Sherwood, A. M., and Sharkey, P. C.,** Immediate effects of spinal cord stimulation, in *Advances in External Control of Human Extremities,* Yugoslav Committee for Electronics and Automation, Belgrade, 1978, 665.
37. **Dimitrijevic, M. R., Dimitrijevic, M. M., Sherwood, A. M., and Gregoric, M. R.,** Neurophysiological study of chronic spinal cord stimulation, in *Advances in External Control of Human Extremities,* Yugoslav Committee for Electronics and Automation, Belgrade, 1978, 673.
38. **Dimitrijevic, M. R., Lehmkuhl, L. D., and Sherwood, A. M.,** Effects of spinal cord stimulation on segmental reflexes in man (abstract), in Proc. Annu. Conf. of the Society for Neuroscience, Atlanta, 1979, 5, 368.
39. **Dimitrijevic, M. R., Faganel, J., Sharkey, P. C., and Sherwood, A. M.,** Study of sensation and muscle twitch responses to spinal cord stimulation, *Int. Rehab. Med.*, 2, 76, 1980.
40. **Dimitrijevic, M. R., Dimitrijevic, M. M., Sherwood, A. M., and Faganel, J.,** Neurophysiological evaluation of chronic spinal cord stimulation in patients with upper motor neuron disorders, *Int. Rehab. Med.*, 2, 82, 1980.
41. **Dimitrijevic, M. R., Sherwood, A. M., Faganel, A. M., and Zidar, J.,** The effect of electrical stimulation of the spinal cord in the phasic and tonic stretch reflexes in extensor muscles of the lower limbs (abstract), in Proc. 28th Int. Congr. of Physiological Sciences, Budapest, 1980.
42. **Dimitrijevic, M. R. and Lenman, J. A. R.,** Neural control of gait in patients with upper motor neurone lesions, in *Spasticity: Disordered Motor Control,* Feldman, R., Young, R., and Koella, W. P., Eds., Symposia Specialists, Miami, 1980, 101.
43. **Dimitrijevic, M. R. and Sherwood, A. M.,** Spasticity: medical and surgical treatment, *Neurology,* 30(2), 19, 1980.
44. **Dimitrijevic, M. R., Faganel, J., Lenman, J. A. R., and Sharkey, P. C.,** Activation of motor units in man by stimulation of anterior and posterior structures of the spinal cord, in Proc. Annu. Conf. Society for Neuroscience, Vol. 6, Cincinnati, 1980, 711.
45. **Dimitrijevic, M. R., Faganel, J., and Young, R. R.,** Underlying mechanisms of the effects of spinal cord stimulation in motor disorders: a review of the discussion, in Proc. 3rd Int. Meeting on Spinal Cord Stimulation, Houston, 1980, *Appl. Neurophysiol.*, 44, 133, 1981.
46. **Dimitrijevic, M. R. and Larsson, L. E.,** Neural control of gait: clinical neurophysiology aspects, in Proc. 3rd Int. Meeting on Spinal Cord Stimulation, Houston, 1980, *Appl. Neurophysiol.*, 44, 152, 1981.
47. **Dimitrijevic, M. R. and Sherwood, A. M.,** A system for epidural spinal cord stimulation, in Proc. 8th Meeting World Soc. Stereotactic and Functional Neurosurgery, Part III, Zurich, 1981, *Appl. Neurophysiol.*, 45, 465, 1982.
48. **Dimitrijevic, M. R., Faganel, J., Sherwood, A. M., and Vodusek, D. B.,** Modification of induced descending spinal cord activation of motor units in man, paper presented at the 12th Annu. Conf. of the Society for Neuroscience, Minneapolis, *Abstracts,* V 8(2):967, 1982.
49. **Dimitrijevic, M. R.,** Neurophysiological evaluation and epidural stimulation in chronic spinal cord injury patients, in *Spinal Cord Reconstruction,* Kao, C., Bunge, R. P., and Reier, P. J., Eds., Raven Press, New York, 1983, 465.
50. **Dimitrijevic, M. R., Faganel, J., and Sherwood, A. M.,** Spinal cord stimulation as a tool for physiological research, *Appl. Neurophysiol.,* 46, 245, 1983.

51. **Dooley, D. M.,** Percutaneous epidural electrical stimulation of the spinal cord, paper presented at the 43rd Annu. Meeting of the American Association of Neurological Surgeons, Bar Harbour, Fla., 1975.
52. **Dooley, D. M.,** Spinal cord stimulation, *A.O.R.N. J.*, 23, 1209, 1976.
53. **Dooley, D. M., Kasprak, M., and Stibitz, M.,** Electrical stimulation of the spinal cord in patients with demyelinating and degenerative diseases of the central nervous system, *J. Fla. Med. Assoc.*, 63, 906, 1976.
54. **Dooley, D. M. and Kasprak, M.,** Modification of blood flow to the extremities by electrical stimulation of the nervous system, *South. Med. J.*, 69, 1309, 1976.
55. **Dooley, D. M. and Sharkey, J.,** Alterations of urinary bladder function produced by electrical stimulation of the spinal cord, *Neurosurgery*, 1(2), 150, 1977.
56. **Dooley, D. M.,** Demyelinating, degenerative, and vascular disease, *Neurosurgery*, 1(2), 220, 1977.
57. **Dooley, D. M. and Sharkey, J.,** Alterations of urinary bladder function produced by electrical stimulation of the spinal cord, paper presented at the 29th Annu. Meeting of the Southern Neurosurgical Society, Innisbrook, Fla., March 3, 1977.
58. **Dooley, D. M. and Sharkey, J.,** Electrostimulation of the nervous system for patients with demyelinating and degenerative diseases of the nervous system and vascular diseases of the extremities, *Appl. Neurophysiol.*, 40, 208, 1977/78.
59. **Dooley, D. M., Sharkey, J., Keller, W., and Kasprak, M.,** Treatment of demyelinating and degenerative diseases by electrostimulation of the spinal cord, in *Advances in External Control of Human Extremities*, Yugoslav Committee for Electronics and Automation, Belgrade, 1978, 529.
60. **Dooley, D. M., Sharkey, J., Keller, W., and Kasprak, M.,** Treatment of demyelinating and degenerative diseases by electrostimulation of the spinal cord, in *Medical Progress Through Technology*, Vol. 6, 1st ed., Springer-Verlag, Basel, 1978, 1.
61. **Dooley, D. M. and Nisonson, I.,** Treatment of patients with degenerative diseases of the central nervous system by electrical stimulation of the spinal cord, in Proc. 3rd Int. Meeting on Spinal Cord Stimulation, Houston, 1980, *Appl. Neurophysiol.*, 44, 71, 1981.
62. Dorsal column stimulation in multiple sclerosis (editorial), *Br. Med. J.*, 6227, 1287, 1980.
63. **Duquette, P., Duquette, J., and Bouvier, G.,** La stimulation electrique de la moelle epiniere dans la sclerose en plaques, *Union Med. Can.*, 109, 890, 1980.
64. **Faganel, J., Dimitrijevic, M. R., Sherwood, A. M., and Zidar, J.,** Electromyographic analysis of muscle twitches evoked by electrical stimulation of the spinal cord, in Abstracts of the 28th Int. Congr. of Physiological Sciences, Budapest, 1980.
65. **Feeney, D. M. and Gold, G. N.,** Chronic dorsal-column stimulation effects on h-reflex and symptoms in a patient with multiple sclerosis, *Neurosurgery*, 6, 564, 1980.
66. **Gildenberg, P. L.,** Treatment of spasmodic torticollis by dorsal column stimulation, *Rev. Inst. Nacion, Neurol.*10, 11, 1976.
67. **Gildenberg, P. L.,** Treatment of torticollis with dorsal column stimulation, *Acta Neurochir.*, 24, 65, 1977.
68. **Gildenberg, P. L.,** Treatment of spasmodic torticollis by dorsal column stimulation, *Appl. Neurophysiol.*, 41, 113, 1978.
69. **Gildenberg, P. L. and Murthy, K. S. K.,** Influence of dorsal column stimulation upon human thalamic somatosensory evoked potentials, *Appl. Neurophysiol.*, 43, 8, 1980.
70. **Gildenberg, P. L.,** Comprehensive management of spasmodic torticollis, in Seminar on Spinal Cord Stimulation, New York, 1980, *Appl. Neurophysiol.*, 44, 233, 1981.
71. **Goodman, S. G. and Young, R. F.,** Dorsal column stimulation in the treatment of multiple sclerosis, *Surg. Forum*, 29, 507, 1978.
72. **Gybels, J. and Van Roost, D.,** Spinal cord stimulation for the modification of dystonic and hyperkinetic conditions: a critical review, in *Recent Advances in Restorative Neurology: Upper Motor Neurone Functions and Dysfunctions*, Eccles, J. C. and Dimitrijevic, M. R., Eds., S. Karger, Basel, in press.
73. **Hawkes, C. H., Wyke, M., Desmond, A., Bultitude, M., Small, D., Jones, S., and Robinson, K.,** Epidural stimulation in 21 patients with multiple sclerosis, in *Advances External Control of Human Extremities*, Yugoslav Committee for Electronics and Automation, Belgrade, 1978, 603.
74. **Hawkes, C. H., Wyke, M., Desmond, A., Bultitude, M., and Kanegaonkar, G. S.,** Stimulation of dorsal column in multiple sclerosis, *Br. Med. J.*, 1, 889, 1980.
75. **Hawkes, C. H.,** Stimulation of dorsal column in multiple schlerosis (letter), *Br. Med. J.* 280, 1619, 1980.
76. **Hawkes, C. H.,** Dorsal column stimulation in multiple sclerosis (letter), *Br. Med. J.*, 281, 147, 1980.
77. **Hawkes, C. H., Fawcett, D., Cooke, E. D., Emson, P. C., Paul, E. A., and Bowcock, S. A.,** Dorsal column stimulation in multiple sclerosis: effects on bladder leg blood flow and peptides, in Proc. 3rd Int. Meeting on Spinal Cord Stimulation, Houston, 1980, *Appl. Neurophysiol.*, 44, 62, 1981.
78. **Illis, L. S., Oygar A. E., Sedgwick, E. M., and Sabbahi Awadalla, M. A.,** Dorsal column stimulation in the rehabilition of patients with multiple sclerosis, *Lancet*, i, 1383, 1976.
79. **Illis, L. S., and Sedgwick, E. M.,** Spinal cord stimulation, *Br. J. Hosp. Med.*, 20, 682, 1978.

80. **Illis, L. S., Sedgwick, E. M., and Tallis. R.,** Evaluation of possible mechanisms of action of spinal cord simulation, in *Advances in External Control of Human Extremities,* Yugoslav Committee for Electronics and Automation, Belgrade, 1978, 647.
81. **Illis, L. S., Sedgwick, E. M., and Tallis, R. C.,** Clinical results of spinal cord stimulation, in *Advances in External Control of Human Extremities,* Yugoslav Committee for Electronics and Automation, Belgrade, 1978, 523.
82. **Illis, L. S., Segwick, E. M., and Tallis, R. C.,** Spinal cord stimulation in neurological disease, *12th Int. Conf. Medical and Biological Engineering,* Jerusalem, August, 1979.
83. **Illis, L. S., Sedgwick, E. M., and Tallis, R. C.,** Spinal cord stimulation in multiple sclerosis: clinical results, *J. Neurol. Neurosurg. Psychiatry.,* 43, 1, 1980.
84. **Illis, L. S. and Sedgwick, E. M.,** Dorsal column stimulation in multiple sclerosis (letter), *Br. Med. J.,* 281, 518, 1980.
85. **Jobling, D. R., Tallis, R. C., Illis, L. S., and Sedgwick, E. M.,** Electronic aspects of spinal cord stimulation, in *Advances in External Control of Human Extremities,* Yugoslav Committee for Electronics and Automation, Belgrade, 1978, 693.
86. **Jobling, D. R., Tallis, R. C., Sedgwick, E. M., and Illis, L. S.,** Electronic aspects of spinal cord stimulation in multiple sclerosis, *Med. Biol. Eng. Comput.,* 10, 40, 1980.
87. **Ketelaer, P., Swartenbroekx, G., Deltenre, P., Carton, H., and Gybels, J.,** Percutaneous epidural dorsal cord stimulation in multiple sclerosis, *Acta Neurochir.,* 49, 95, 1979.
88. **Kiwerski, J.,** Stimulation of the spinal cord after its traumatic injury, in Proc. Int. Symp. Acute Management and Rehabilitation After Spinal Cord Injuries, Metropolitan Center for Rehabilitation, Warsaw-Konstancin, 1977, 207.
89. **Kiwerski, J., Weiss, M., and Pasniczek, R.,** Clinical results in traumatic spinal cord injury electrical stimulation, in *Advances in External Control of Human Extremities,* Yugoslav Committee for Electronics and Automation, Belgrade, 1978, 169.
90. **Klinger, D. and Kepplinger, B.,** Quantification of the effects of epidural spinal electrostimulation (E.S.E.S.) in central motor disorders, in Proc. 8th Meeting World Soc. Stereotatic and Functional Neurosurgery, Part I, Zurich, 1981, *Appl. Neurophysiol.,* 45, 221, 1982.
91. **Klinger, D., Kepplinger, B., Gerstenbrand, F., and Heinisch, W.,** Die epidurale spinale Elektrostimulation E.S.E.S. bei chronischen Schmerzustanden und zentralen motorischen Storungen, *Wien. Klin. Wochenschr.,* in press.
92. **Krainick, J. U., Thoden, U., Strassburg, H. M., and Wenzel, D.,** The effect of electrical spinal cord stimulation on spastic movement disorders, in *Advances in Neurosurgery,* Vol. 4, Wullenweber, R., et al., Eds., Springer-Verlag, Berlin, 1977, 257.
93. **Laitinen, L. V. and Fugl-Meyer, A. R.,** Assessment of functional effect of epidural electrostimulation and selective posterior rhizotomy in spasticity, in Proc. 8th Meeting World Soc. Stereotatic and Functional Neurosurgery, Part I, Zurich, 1981, *Appl. Neurophysiol.,* 45, 331, 1982.
94. **Lazorthes, Y., Siegfried, J., and Broggi, G.,** Electrical spinal cord stimulation for spastic motor disorders in demyelinating diseases — a cooperative study, in *Indications for Spinal Cord Stimulation,* Hosobuchi, Y. and Corbin, T., Eds., Exerpta Medica, New York, 1981, 48.
95. **Levine, B. E. and Hubschmann, O. R.,** Dorsal column stimulation: effect on human cerebrospinal fluid and plasma catecholamines, *Neurology,* 30, 65, 1980.
96. **Levita, E., Rilan, M., and Waltz, J. M.,** Psychological effects of spinal cord stimulation: preliminary findings, in Proc. 3rd Int. Meeting on Spinal Cord Stimulation, Houston, 1980, *Appl. Neurophysiol.,* 44, 93, 1981.
97. **Lortie-Monette, F.,** Dorsal column stimulation in multiple sclerosis (letter), *Br. Med. J.,* 281, 457, 1980.
98. **Meglio, M., Cioni, B., D'Amico, E., Ronzoni, G., and Rossi, G. F.,** Epidural spinal cord stimulation for the treatment of neurogenic bladder, *Acta Neurochir.,* 54, 191, 1980.
99. **Murthy, K. S. K.,** Propriospinal mechanisms involved in the effects of spinal cord stimulation, in Proc. 3rd Int. Meeting on Spinal Cord Stimulation, Houston, 1980, *Appl. Neurophysiol.* , 44, 22, 1981.
100. **Nashold, B. S., Jr.,** Overview of neuroaugmentation, *Neurosurgery,* 1, 230, 1977.
101. **Phillips, C. G.,** Possible modes of action of extradural electrical stimulation on the spinal cord, in Proc. 3rd Int. Meeting on Spinal Cord Stimulation, Houston, 1980, *Appl. Neurophysiol.,* 44, 16, 1981.
102. **Plotkin, R.,** Pisces stimulation for motor neurone disease, *Acta Neurochir. Suppl.,* 30, 429, 1980.
103. **Read, D. J., Matthews, W. B., and Higson, R. H.,** The effect of spinal cord stimulation on function in patients with multiple sclerosis, *Brain,* 103, 803, 1980.
104. **Reynolds, A. F. and Oakley, J. C.,** High frequency cervical epidural stimulation for spasticity, in Proc. 8th Meeting World Soc. Stereotatic and Functional Neurosurgery, Part I, Zurich, 1981, *Appl. Neurophysiol.,* 45, 93, 1982.
105. **Richardson, R. R. and McLone, D. G.,** Percutaneous epidural neurostimulation for paraplegic spasticity, *Surg. Neurol.,* 9, 153, 1978.

106. **Richardson, R. R., Cerullo, L. J., and Meyer, P. R.,** Autonomic hyper-reflexia modulated by percutaneous epidural neurostimulation. A preliminary report, *Neurosurgery*, 4, 517, 1979.
107. **Richardson, R. R., Cerullo, L. J., McLone, D. G., Gutierrez, F. A., and Lewis, V.,** Percutaneous epidural neurostimulation in modulation of paraplegic spasticity. Six case reports, *Acta Neurochir.*, 49, 235, 1979.
108. **Rosen, J. A. and Barosum, A. H.,** Failure of chronic dorsal column stimulation in multiple sclerosis, *Ann. Neurol.*, 6, 66, 1979.
109. **Roohi, F.,** Spinal cord stimulation and motor unit action potentials, in *Advances in External Control of Human Extremities*, Yugoslav Committee for Electronics and Automation, Belgrade, 1978, 611.
110. **Scerrati, M., Onofrj, M., and Pola, P.,** Effects of spinal cord stimulation on spasticity: H-reflex study, in Proc. 8th Meeting World Soc. Stereotatic and Functional Neurosurgery, Part I, Zurich, 1981, *Appl. Neurophysiol.*, 45, 62, 1982.
111. **Sedan, R. and Lazorthes, Y.,** La neurostimulation dans les syndromes spastiques et moteurs, *Neurochirurge*, 24 (Suppl. 1), 81, 1978.
112. **Sedgwick, E. M., Thornton, A. R. D., El-Negamy, E., Tallis, R. C., and Illis, L. S.,** Electrophysiological responses associated with spinal cord stimulation, in *Advances in External Control of Human Extremities*, Yugoslav Committee for Electronics and Automation, Belgrade, 1978, 635.
113. **Sedgwick, E. M., Illis, L., Tallis, R. C., Thornton, A. R. D., Abraham, P., El-Negamy, E., Docherty, R. B., Soar, J. S., Spencer, S. C., and Taylor, F. M.,** Evoked potentials and contingent negative variation during treatment of multiple sclerosis with spinal cord stimulation, *J. Neurol. Neurosurg. Psychiatry*, 43, 15, 1980.
114. **Sedgwick, E. M., Thornton, A. R. D., El-Negamy, E., Jobling, D. R., Tallis, M. B., and Illis, L. S.,** Electrophysiological responses and stimulus parameters associated with spinal cord stimulation, *Int. Rehab. Med.*, 2, 71, 1980.
115. **Sharkey, P. C., Dimitrijevic, M. M., and Campos, R. J.,** Effects of spinal cord stimulation on modification of motor disability in patients with upper motor neurone disorders, in *Proc. 5th Ann. Conf. Systems and Devices for the Disabled*, The Institute for Rehabilitation and Research, Houston, 1978, 205.
116. **Sharkey, P. C., Dimitrijevic, M. M., Campos, R. J., and Carter, R. E.,** Effects of spinal cord stimulation on spasticity in patients with sustained spinal cord injury, in Proc. 6th Annu. Scientific Meeting of the American Spinal Injury Association, Anaheim, Calif., 1980, 35.
117. **Sharkey, P. C., Dimitrijevic, M. M., Campos, R. J., and Dimitrijevic, M. R.,** Remarks on spinal cord stimulation and the placebo effect, in Proc. 3rd Int. Meeting on Spinal Cord Stimulation, Houston, 1980, *Appl. Neurophysiol.*, 44, 119, 1981.
118. **Sharkey, P. C.** Technological problems of spinal cord stimulation systems: a clinical perspective, in Proc. 3rd Int. Meeting on Spinal Cord Stimulation, Houston, 1980, *Appl. Neurophysiol.*, 44, 50, 1981.
119. **Sharkey, P. C., Dimitrijevic, M. M., and Faganel J.,** Neurophysiological analysis of factors influencing efficacy of spinal cord stimulation, *Appl. Neurophysiol.*, 45, 68, 11982.
120. **Sherwood, A. M., Sharkey, P. C., Canzoneri, J., and Dimitrijevic, M. R.,** Biomedical engineering specifications for epidural spinal cord stimulation to augment motor performance, in *Advances in External Control of Human Extremities*, Yugoslav Committee for Electronics and Automation, Belgrade, 1978, 703.
121. **Sherwood, A. M., Sharkey, P. C., Dimitrijevic, M. R., Dimitrijevic, M. M., and Campos, R. J.,** Spinal cord stimulation for modification of motor disability in patients with upper motor neurone diseases, in Proc. 31st Annu. Conf. on Engineering in Medicine and Biology, Atlanta, 20, 275, 1978.
122. **Sherwood, A. M., Sharkey, P. C., and Dimitrijnevic, M. R.,** Techniques for epidural spinal cord stimulation, in *Proc. 5th Annu. Conf. Systems and Devices for the Disabled*, The Institute for Rehabilitation and Research, Houston, 1978, 210.
123. **Sherwood, A. M., Sharkey, P. C., and Dimitrijevic, M. R.,** Biomedical engineering specifications for epidural spinal cord stimulation to augment motor performance, *Int. Rehab. Med.*, 2, 62, 1980.
124. **Sherwood, A. M.,** Biomedical engineering aspects of spinal cord stimulation, in Proc. 3rd Int. Meeting on Spinal Cord Stimulation, Houston, 1980, *Appl. Neurophysiol.*, 44, 126, 1981.
125. **Siegfried, J.,** Traitement neurochirurgical de quelques symptomes de la sclerose en plaques, *Med. Hyg.*, 34, 901, 1976.
126. **Siegfried, J.,** Le role de la neurochirurgie fonctionnelle dans le traitement du tremblement intentionnel, del la nevralgie faciale et de la spasticite de la sclerose en plaques, *Oto-Neuro-Opthal.*, 48, 359, 1976.
127. **Siegfried, J.,** Traitment neurochirurgical de l'hypertonie musculaire, *Med. Hyg.*, 35, 2062, 1977.
128. **Siegfried, J.,** Deux aspects differents du traitement neurochirurugical de la spasticite: interventions cerebrales et stimulation des cordons posterieurs de la moelle, *Neurochirurgie*, 23, 344, 1977.
129. **Siegfried, J., Krainick, J. U., Haas, H., Adorjani, C., Meyer, M., and Thoden U.,** Electrical spinal cord stimulation for spastic movement disorders, *Appl. Neurophysiol.*, 41, 134, 1978.
130. **Siegfried, J.,** Treatment of spasticity by dorsal column stimulation, *Int. Rehab. Med.*, 2, 31, 1980.

131. **Siegfried, J., Lazorthes, Y., and Broggi, G.,** Electrical spinal cord stimulation for spastic movement disorders, in Proc. 3rd Int. Meeting on Spinal Cord Stimulation, Houston, 1980, *Appl. Neurophysiol.*, 44, 77, 1981.
132. **Spiegel, E. A.,** Relief of pain and spasticity by posterior column stimulation. A proposed mechanism, *Arch. Neurol.*, 39, 184, 1982.
133. **Tallis, R. C., Illis, L. S., and Sedgwick, E. M.,** Urodynamic changes with spinal cord stimulation, in *Advances in External Control of Human Extremities*, Yugoslav Committee for Electronics and Automation, Belgrade, 1978, 575.
134. **Thoden, U., Krainick, J. U., Strassburg, H. M., and Wenzel, D.,** Modulation of monosynaptic reflexes by dorsal column stimulation (D.C.S) in spasticity, *Eur. J. Physiol.*, 365, 41, 1976.
135. **Thoden, U., Krainick, J. U., Strassburg, H. M., and Zimmermann, H.,** Influence of dorsal column stimulation (D.C.S) on spastic movement disorders, *Acta Neurochir.* 39, 233, 1977.
136. **Vodovnik, L.,** Indirect spinal cord stimulation. Some engineering viewpoints, in Proc. 3rd Int. Meeting on Spinal Cord Stimulation, Houston, 1980, *Appl. Neurophysiol.*, 44, 97, 1981.
137. **Waldron, J. and Tallis, R. C.,** Spinal cord stimulation in the treatment of multiple sclerosis, *Radiography*, 65, 114, 1979.
138. **Waltz, J. M. and Pani, K. C.,** Spinal cord stimulation in disorders of the motor system, in *Advances in External Control of Human Extremities*, Yugoslav Committee for Electronics and Automation, Belgrade, 1978, 545.
139. **Waltz, J. M.,** Spinal cord stimulation for palsies?, *Patient Care*, 13, 118, 1979.
140. **Waltz, J. M. and Andreesen, W. H.,** Multiple-lead spinal cord stimulation: technique, in Proc. 3rd Int. Meeting on Spinal Cord Stimulation, Houston, 1980, *Appl. Neurophysiol.*, 44, 30, 1981.
141. **Waltz, J. M., Reynolds, L. O., and Riklan, M.,** Multi-lead spinal cord stimulation for control of motor disorders, Seminar on Spinal Cord Stimulation, New York, 1980, *Appl. Neurophysiol.*, 44, 244, 1981.
142. **Waltz, J. M.,** Computerized percutaneous multi-level spinal cord stimulation in motor disorders, in Proc. 8th Meeting World Soc. Stereotatic and Functional Neurosurgery, Part I, Zurich, 1981, *Appl. Neurophysiol.*, 45, 73, 1982.
143. **Young, R. F. and Goodman, S. J.,** Dorsal spinal cord stimulation in the treatment of multiple sclerosis, *Neurosurgery*, 5, 225, 1979.
144. **Bamford, C., et al.,** Anesthesia in multiple sclerosis, *Can. J. Neurol. Sci.* 5, 41, 1978.
145. **Baskett, P. J. and Armstrong, R.,** Anaesthetic problems in multiple sclerosis, are certain agents contraindicated?, *Anaesthesia*, 25, 397, 1970.
146. **Bergveld, P.,** Simple test for electrophysiologically tolerable parameters of artificial stimulation, *Med. Biol. Eng.*, 14, 497, 1976.
147. **Brummer, S. B. and Turner, M. J.,** Electrochemical considerations for safe electrical stimulation of the nervous system with platinum electrodes, *IEEE Trans. Biomed. Eng.*, 24, 59, 1977.
148. **Canosa, F. L., Dooley, D. M., and Boyle, E.,**Vascular reactions in pre- and post-nerve root and spinal cord transcutaneous electrical stimulation in *Advances in External Control of Human Extremities*, Yugoslav Committee for Electronics and Automation, Belgrade, 1978, 585.
149. **Coburn, B.,** Electrical stimulation of the spinal cord: two-dimensional finite element analysis with particular reference to epidural electrodes, *Med. Biol. Eng. Comp.*, 18, 573, 1980.
150. **Coburn, B.,** A Theoretical Analysis of Spinal Cord Stimulation with Particular Reference to Epidural Electrodes, Ph.D. thesis, Brunel University, Uxbridge, Middlesex, England, 1981.
151. **Coburn, B.,** Finite element analysis of stimulation fields produced by epidural stimulation, paper presented at the 34th Annu. Conf. Engineering in Medicine and Biology, Houston, September 22, 1981.
152. **Comte, P.,** Monopolar versus bipolar stimulation, in Proc. 8th Meeting World Soc. Stereotatic and Functional Neurosurgery, Part I, Zurich, 1981, *Appl. Neurophysiol.*, 45, 156, 1982.
153. **Davis, F.,** Pathophysiology of multiple sclerosis and related clinical implications, *Mod. Treatment*, 7, 890, 1970.
154. **Dimitrijevic, M. R., Spencer, W. A., Trontelj, J. V., and Dimitrijevic, M. M.,** Reflex effects of vibration in patients with spinal cord lesions, *Neurology*, 27(11), 1078, 1977.
155. **Dimitrijevic, M. R., Nathan, P. W., and Sherwood, A. M.,** Clonus: the role of central mechanisms, *J. Neurol. Neurosurg. Psychiatry*, 43, 321, 1980.
156. **Dimitrijevic, M. R., Faganel, J., Lehmkuhl, D., and Sherwood, A. M.,** Motor control in man with spinal cord injury, in *Brain and Spinal Mechanisms of Movement Control in Man: New Developments and Clinical Applications*, Desmedt, J. E., Ed., Raven Press, New York, 1983, 915.
157. **Dimitrijevic, M. R., Lenman, J., Prevec, T., and Wheatly, K.,** A study of posterior column function in familial spastic paraplegia, *J. Neurol. Neurosurg. Psychiatry*, 45, 46, 1982.
158. **Ertekin, C.,** Human evoked electrospinogram, in *New Developments in Electromyography and Clinical Neurophysiology*, Vol. 2, Desmedt, J. E., Eds., S. Karger, Basel, 1973, 344.
159. **Ertekin, C., Yakup, S., and Uckardesler, L.,** Studies on the human spontaneous electromyelogram. I. Normal subjects, *Electroencephalogr. Clin. Neurophysiol.*, 55, 13, 1983.

160. **Faganel, J. and Dimitrijevic, M. R.,** Study of propriospinal interneurone system in man: cutaneous exteroceptive conditioning of stretch reflexes, *J. Neurol. Sci.*, 56, 155, 1982.
161. **Frolich, A. and Sherrington, C.,** Path of impulses for inhibition under decerebrate rigidity, *J. Physiol. (London)*, 28, 14, 1902.
162. **Geddes, L. A. and Baker, L. E.,** The specific resistance of biological material — a compendium of data for the biomedical engineer and physiologist, *Med. Biol. Eng.*, 5, 271, 1967.
163. **Janko, M. and Trontelj, H. V.,** Flexion withdrawal reflex as recorded from single human biceps femoris motor neurones, *Pain*, 15, 167, 1983.
164. **Kakulas, B. A. and Bedbrook, G. M.,** A correlative clinico-pathologic study of spinal cord injury, *Proc. Aust. Assoc. Neurol.*, 6, 123, 1969.
165. **Larson, S. J., Sances, A., Cusick, J. F., Meyer, G. A., and Swiontek, T. J.,** A comparison between anterior and posterior spinal implant systems, *Surg. Neurol.*, 4, 180, 1975.
166. **Larson, S. J., Sances, A., Riegel, D. H., Meyer, G. A., Dallman, D. E., and Swiontek, T. J.,** Neurophysiological effects of dorsal column stimulation in man and monkey, *J. Neurosurg.*, 41, 217, 1974.
167. **Law, J. D. and Miller, L. V.,** Importance and documentation of an epidural stimulating position, in Proc. 8th Meeting World Soc. Stereotatic and Functional Neurosurgery, Part III, Zurich, 1981, *Appl. Neurophysiol.*, 45, 461, 1982.
168. **Lilly, J. C.,** Injury and excitation by electric currents. A. The balanced pulse-pair waveform, in *Electrical Stimulation of the Brain*, Sheer, D. E., Ed., University of Texas Press, Austin, 1961, 60.
169. **Long, D. M. and Hagfors, N.,** Electrical stimulation of the nervous system: the current status of electrical stimulation of the nervous system for relief of pain, *Pain*, 1, 109, 1975.
170. **Lundberg, A.,** Supraspinal control of transmission in reflex paths to motoneurones and primary afferents, *Prog. Brain Res.*, 12, 197, 1964.
171. **McNeal, D. R.,** Analyses of a model for excitation of myelinated nerve, *IEEE Trans. Biomed. Eng.*, 23, 329, 1976.
172. **Melzack, R. and Wall, P. D.,** Pain mechanisms: a new theory, *Science*, 150, 971, 1965.
173. **Nashold, B. S., Grimes, J., Friedman, H., Semans, J., and Avery, R.,** Operative stimulation of the neurogenic bladder, *Neurosurgery*, 1(2), 218, 1977.
174. **Pineda, A.,** Complications of dorsal column stimulation, *J. Neurosurg.*, 48, 64, 1978.
175. **Pudenz, R. H.,** Adverse effects of electrical energy applied to the nervous system, *Neurosurgery*, 1, 190, 1977.
176. **Rusinko, J. B., Walker, C. F., and Sepulveda, N. G.,** Finite element modeling of potentials within the human thoracic spinal cord due to applied electrical stimulation, in *Proc. 3rd Ann. Conf. Engineering in Medicine and Biology Group*, Frontiers of Engineering in Health Care — 1981, IEEE, Piscataway, N.J., 1981, 76.
177. **Shealy, C. N., Mortimer, J. T., and Reswick, J. B.,** Electrical inhibition of pain by dorsal column stimulation: preliminary clinical report, *Anaesth. Analg.*, 46, 489, 1967.
178. **Sherwood, A. M.,** Characteristics of evoked potentials recorded over the spinal cord and brain of man, *IEEE Trans. Biomed. Eng.*, 28, 481, 1981.
179. **Sherwood, A. M.,** Somatosensory evoked potentials recorded in the epidural space in man, in Frontiers of Engineering in Health Care — 1982, *Proc. 4th Ann. Conf. Engineering in Medicine and Biology*, IEEE, Piscataway, N.J., 1982, 50.
180. **Sherwood, A. M.,** E.M.G of upper motor neurone dysfunction, paper presented at the *IEEE Region V Conf.*, Houston, April 20, 1983, IEEE, Piscataway, N.J., 1983, 58.
181. **Shimoji, K., Kano, T., Higashi, H., and Kano, T.,** Epidural recording of spinal electrogram in man, *Electroencephalogr. Clin. Neurophysiol.*, 30, 236, 1971.
182. **Shimoji, K., Kano, T., Higashi, H., Morioka, T., and Henschel, E. O.,** Evoked spinal electrograms recorded from epidural space in man, *Appl. Neurophysiol.*, 33, 468, 1972.
183. **Shimoji, K., Kano, T., Morioka, T., and Ikezono, E.,** Evoked spinal electrogram in a quadriplegic patient, *Electroencephalogr. Clin. Neurophysiol.*, 35, 659, 1973.
184. **Smith, C.,** Instrumentation problems, Symposium on Neuroaugmentive Devices, *Neurosurgery*, 1, 224, 1977.
185. **Swiontek, T. J.,** Distribution of Applied Electrical Current in the Spinal Cord and its Effect on Evoked Potentials, Ph.D. thesis, Marquette University, Milwaukee, 1975.
186. **Swiontek, T. J., Sances, A., Larson, S. J., Ackmann, J. J., Cusik, J. F., Meyer, G. A., and Millar, E. A.,** Spinal cord implant studies, *IEEE Trans. Biomed. Eng.*, 23, 307, 1976.
187. **Wall, P. D.,** The sensory and motor role of impulses traveling in the dorsal columns towards cerebral cortex, *Brain*, 93, 505, 1970.
188. **Wall, P. D. and Noordenbos, W.,** Sensory functions which remain in man after complete transection of dorsal columns, *Brain*, 100, 641, 1977.
189. **Zienkewicz, O. C.,** *The Finite Element Method*, 3rd ed., McGraw-Hill, New York, 1977.

Chapter 8

SPINAL CORD STIMULATION FOR PAIN

Dennis J. Maiman, Sanford J. Larson, and Anthony Sances, Jr.

TABLE OF CONTENTS

I. INTRODUCTION

Since the initial report of Shealy et al.[45] more than 10,000 patients have had electrodes implanted on the spinal cord for pain management. There are more than 2000 spinal stimulators implanted each year in the U.S. and an additional 3400 are implanted in the rest of the world.[9] Over the last 10 years, the number of implants performed each year has grown at an annual rate of 20%. Additionally, approximately 1000 more have been utilized in the management of motor disorders.[4,33,34,37,43]

In the early series, electrode arrays were placed subdurally over the dorsal columns following laminectomy. Initial pain relief was observed in most patients; however, the applied voltages from the stimulating devices usually had to be increased with time. The paresthesias also produced were often more unpleasant than the pain itself.[16] Because of these, techniques have been developed to apply the currents to the ventral aspect of the cord, with the currents then extending to the dorsal columns.[10,17,41] More recently, the majority of investigators have utilized percutaneously inserted epidural electrodes for the management of chronic pain.[5,14,20-24,30,31,38,46,51,54] With this method and improved patient selection, long-term pain relief has been obtained in approximately 50% of the implanted subjects.

The original studies utilizing spinal cord and peripheral nerve stimulation for pain were based upon the gate control theory developed by Melzack and Wall.[25] Briefly, it was proposed that the transmission of information related to noxious stimuli is controlled at spinal cord levels; the degree of transmission is determined by the balance of activity between large- and small-fiber peripheral systems. It was suggested that activation of large fibers would inhibit the relay of impulses from small fibers. Subsequently, Wall and Sweet[52] obtained pain relief by stimulation of peripheral nerves and Shealy et al.[45] applied currents to the dorsal columns for intractable pain. Although the gate control theory has been criticized as not accounting for much of the physiological evidence available,[29] it continues to be advanced as an explanation for pain perception and the effectiveness of peripheral and spinal cord stimulation for pain.[53] Experimental support for the theory was reported by Foreman et al., who demonstrated suppression of spinothalamic neurons during activation of the dorsal columns.[7] However, Nathan and Rudge were unable to elevate pain thresholds with selective stimulation of peripheral large fibers.[28] Several investigators could not obtain clinical pain relief with currents confined to the dorsal columns,[2,8] and Bantli et al.[1] suggested that the amplitudes used for clinical stimulation are sufficient to activate most of the long tracts of the spinal cord.

Larson et al.[17] and Hoppenstein[10] observed suppression of pain at lower stimulation levels when the electrodes were placed anterior to the spinal cord and hypothesized that the effect is due to a conduction block in the spinothalamic tract. Campbell, on the other hand, suggested that Lissauer's tract is the site of action.[2] Although it is generally assumed that the effects of applying current to the spinal cord are due to the activation of fibers,[53] the characteristics of the electrophysiologic alterations are not inconsistent with a depolarizing block preferentially in the small fibers.[16,17]

Although activation of dorsal column axons may not be the explanation for relief of pain following application of current, it is clear that dorsal column stimulation works. Certainly, it is much simpler to implant electrodes over the dorsal columns than, for example, anterior to the spinal cord. However, when pain is located in the perineum and lower limbs, our patients have reported greater relief of pain when currents are applied through electrodes placed anterior to the spinal cord.

II. IMPLANT METHODS

Implantable units are manufactured by several companies. Although these units differ with respect to electrode configuration, they have certain features in common. Each is of unitary construction with a small s.c. receiver connected by cable to the electrode set. The currents are transmitted through the skin, picked up by a passive receiver implanted subcutaneously, and delivered to the cord. As the level of applied current is increased, the patient notices paresthesias below the level at which the electrodes were placed. Some patients have developed sensory levels with ankle clonus, increased deep reflexes, and pathologic reflexes. These findings disappear after the currents are shut off. When currents are applied through anterior electrodes, paresthesias are not experienced, although perception of pain is diminished.

The surgical technique for implantation of subdural electrodes is not complicated and has been described in detail elsewhere.[16,17] The electrode sets are relatively small and narrow, and an extensive laminectomy is not required. For example, if it is desired to place the implant at the T7-T8 level, it is only necessary to remove the inferior portion of the spinous process and lamina of T7 and superior portion of T8.

For dorsal column implants, we use a set of 5-in-line 2-mm-diameter platinum-iridium disc electrodes embedded in a thin sheet of silastic, measuring 20 × 6 mm. Electrodes 1, 3, and 5 are connected to the positive terminal, and 2 and 4 are connected to the negative. A silastic keel is attached to the sheet on the side opposite the electrodes. The keel serves to support the central electrode and also to facilitate fixation of the unit to the dura.

For anterior implants we use the 5-in-line electrode array, but without a keel. The cable is incorporated in the silastic sheet in which the electrodes are embedded and is oriented at right angles to the electrode array. The implant is autoclaved around a test tube 1 cm in diameter to provide approximate conformity to the contour of the spinal cord. A wide laminectomy is required, with exposure of the nerve root sheaths bilaterally above and below the level of implant. A two-level rhizotomy is necessary because the amount of current that can be delivered may be limited by muscle contractions induced by electrical stimulation of these roots.

To check the integrity of the system prior to closure, electric stimuli are applied at low frequency through the electrode set and evoked potentials are recorded from scalp electrodes. Following surgery, the implants are evaluated by monitoring the effect of high-frequency spinal stimulation on the scalp responses evoked by peripheral nerve stimulation.[17,26,27,40]

It is generally possible to begin transcutaneous transmission of current to the receiver within a few days following operation. The frequency of the currents applied by the transmitter may be varied between 5 and 200 Hz, with pulse current amplitudes up to 3 mA and durations of 0.1 to 1 msec. Capacitative coupling is used to minimize tissue damage.[47,48] In most patients, frequencies of 70 to 100 Hz are used with pulse widths of 0.25 msec and an estimated pulse current of 0.5 to 1 mA. The procedure for subdural spinal implant is not lengthy, entails minimal blood loss and tissue trauma, and if necessary, can be performed under local anesthesia.

The percutaneous implantation of epidural electrodes has become the more utilized technique in recent years. The procedure, as described by LeRoy et al.[18,19] and Long et al.,[20-24] is generally performed with fluoroscopy and under local anesthesia. A small incision is made between two upper lumbar spinous processes. Touhy needles are then used to pierce the interspinous ligament and enter the epidural space. The wire electrodes are then passed cephalad through the needles, staying as close to the midline as possible. Many investigators place one electrode one level higher than the other.

Intraoperative testing is generally carried out. The electrodes are positioned so that during stimulation maximum paresthesias in the painful regions are created. Polarities, frequencies, and current levels may be altered to determine the most beneficial parameters.

Stimulation is usually performed regularly for the first few postoperative days. If adequate analgesia is produced over this period of time, the system is internalized. Typically utilized current parameters are similar to those discussed above.

III. CLINICAL STUDIES OF SPINAL CORD STIMULATION

Patients underwent spinal cord stimulation using the subdural method described above. They were divided into two groups: anterior and posterior electrode placement. The level of current required for reduction of the evoked potential amplitude and the production of paresthesias was measured by recording the pulse current at each of the five electrodes.

In the patients with dorsal column units, the intensity of the applied currents was gradually increased until marked or complete pain relief was achieved, which occurred in nearly all patients. Each patient described paresthesias proportionate to current levels, which to some were as unpleasant as the pain. In most patients the initial good results were maintained until death (6 weeks to 8 months after implant). However, in approximately one third of the patients, decreasing analgesia was seen 1 to 3 months postoperatively.

Relief of pain usually persisted for several hours after each period of stimulation. In two thirds of the patients, decreased perception of joint rotation, pain, and touch, plus increased deep reflexes, some pathological reflexes, and ankle clonus, were observed below the level of current application. Somatosensory-evoked potential amplitudes were markedly reduced in these patients during current application. All neurological findings and evoked potentials returned to control values within 30 to 60 min after current application.

The findings in two patients deserve special mention. One was a 49-year-old man with carcinoma of the pancreas and severe abdominal pain. Electrodes were implanted at the level of the seventh thoracic vertebra. When currents were applied, he developed reduced sensory perception below T8 with corresponding relief of pain. However, he continued to have a small area of pain immediately above the sensory level.

The other patient was a 53-year-old man with carcinoma of the kidney and pelvic metastases. Initially, he complained of severe lower abdominal pain. The electrodes were implanted at about T8. When currents were applied, he obtained excellent relief of pain with decreased perception of pain, touch, and joint rotation below the level of the umbilicus. However, patellar and ankle reflexes became hyperactive, and he developed ankle clonus and extensor toe signs. Cortical potentials evoked by peroneal nerve stimulation were markedly reduced in amplitude. All neurological findings and evoked potential amplitudes returned to control levels 30 min after the currents were stopped, although pain relief persisted for several hours. He continued to obtain good relief from pain until the unit abruptly stopped working about 4 months after operation. Before it could be replaced, he developed pain in the right shoulder and upper right arm secondary to apical metastases. A laminectomy of C6 was done and the electrodes implanted subdurally. A higher cord level was avoided because of concern over possible interference with pulmonary function. With lower levels of applied current the patient noticed paresthesias in the left leg. As the intensity was increased, he noticed paresthesias in the right leg, then in the left side of the body, the left arm and shoulder, and eventually the right arm and the right shoulder. Perception of touch, pin-prick, and pain was reduced between C7 and T12. The abdominal pain was relieved, but the pain in the right shoulder and arm was only minimally affected. Perception of joint rotation remained normal in the thumbs, was somewhat decreased in the index fingers, and more markedly decreased in the middle, ring, and small fingers. Potentials evoked by digital nerve stimulation and recorded from scalp electrodes were correspondingly affected, remaining normal from the thumb, but markedly reduced in amplitude from the other fingers. The patient subsequently began to develop increasingly severe pain in the left hip. Although he felt paresthesias in this area when the currents were applied, the pain was not relieved. Again, perception of pain and touch was normal below T12. With a substantial increase in current intensity producing paresthesias described as nearly intolerable, the patient began to notice a substantial, although incomplete, relief of hip pain. Reexamination demonstrated that the lower level of hypalgesia was now at L2. The pain returned promptly when the currents were stopped.

Six patients were implanted with electrodes over the anterior aspect of the spinal cord as well as the usual dorsal column system.[17,41,47] With stimulation of the ventral electrodes, pain relief was obtained without the paresthesias or somatosensory-evoked potential changes usually noted with dorsal column stimulation. With increased currents, painful muscle contractions were sometimes observed. The use of low-level dorsal column stimulation, in addition to anterior stimulation, produced pain relief at currents below those required for muscle contractions and paresthesias.

When currents were passed from anterior to posterior in the spinal cord, pain relief was obtained at significantly lower current levels. The best effect was noted when the anterior electrodes were negative while the posterior electrodes were positive. Pain relief was obtained without paresthesias and the somatosensory-evoked potential was unaffected.[41,42]

In four patients with implants, the portion of the spinal cord beneath the electrode set was removed at autopsy. Each of these patients had had current applied to the cord for a total of at least 100 hr. Spinal cords were also otained from four patients who died of cancer, but had no evidence of metastases involving the CNS, and from four patients who died of nonneoplastic disorders. The spinal cords were stained by the Marchi method to demonstrate degenerating myelin. Microscopic examination did not demonstrate any evidence of lesions beneath the electrodes. Scattered degeneration was seen in three of four patients with implants, three of four cancer patients without implants, and four of four patients who died of nonneoplastic disease.

IV. DISCUSSION

The findings from these and other experiments indicate that application of currents to the spinal cord at high frequency may produce a transmission block in that portion of the spinal cord beneath the electrodes. As would be expected, the fibers closest to the electrode set are more profoundly affected than more distant axons. This appears to be a depolarizing block, since negative electrodes are more efficient than positive. The persistence of transmission block after the currents have been turned off suggests a chemical as well as an electrical effect.

With the techniques described, pain relief in patients appears related to interference with transmission over the fibers of the spinothalamic tract. When transmission in the spinothalamic tract is affected by currents applied through electrodes placed on the dorsal columns, the current density in the dorsal columns must necessarily be substantially higher than the current density in the anterior quadrants. This probably accounts for the difficulty encountered in relieving pain in the lower lumbar and sacral dermatomes, since the paresthesias become intolerable before an effective current density is obtained in the more lateral portions of the spinothalamic tract. However, application of current through anterior electrodes will produce analgesia without paresthesias below the level of implant, when discomfort from nerve root stimulation at electrode levels does not limit the amount of current used.

Other electrode configurations and methods of current application may prove advantageous. An array of two parallel sets of five-in-line electrodes located beneath the anterior quadrants on either side of the midline would tend to provide an effective current density in the anterior quadrants, with less current applied to each electrode than is necessary with the midline array.

The use of anterior electrodes alone in the cervical region does not seem practical because of the extensive rhizotomy which would probably be required. However, by passing currents across the cervical spinal cord using electrodes placed on the posterior and anterior surfaces, the current density is greatest in the midsagittal portion of the spinal cord. Teledeltos paper plots indicated that 50% of the current applied between the anterior and posterior electrodes is retained in a parasagittal zone extending 3 mm on either side of the midline. A comparative study of current distribution in human cadaver cords, saline tank model, and Teledeltos paper suggested that this electrode system confines the current to the midsagittal portion of the cord.[16,17,26,41,47,49] Consequently, neural transmission could be blocked in the medial portion of the spinothalamic tract, producing a dissociated sensory loss in the upper limbs without significant muscle contraction.

Implantation of this type of unit in the lower thoracic area could be used for control of pain in the perineum and lower limbs. However, since the sacral representation in the spinothalamic tract is more lateral, problems may be encountered because the currents may also affect the corticospinal tract.

Because the positive and negative electrodes are in separate pads, passing current through the cord from anterior to posterior produces a current distribution which is not disturbed by thickening of the arachnoid beneath the electrodes, since current shunting cannot occur. Therefore, more consistent long-term results would be anticipated.

Epidural monopolar electrodes currently in use inject less current into the spinal cord than the alternate polarity arrays or the anterior to posterior array.[3,39,44,49] However, they probably produce relatively uniform current densities in the spinal cord. The fact that the small-fiber systems are blocked at lower current levels than large-fiber systems contributes to the effectiveness of these units.[50] Furthermore, since the electrode and current return paths are separate, shunting of currents due to electrode encapsulation does not occur, and units do not require higher currents over time. Regardless of technique and mechanisms, several salient points become apparent when looking at most investigators' series.

The importance of careful patient selection has been repeatedly emphasized. When North et al. evaluated their long-term results with spinal cord stimulation, they found marked improvement in results with more selective criteria for implantation.[30,31] The recommended criteria include psychological testing and intraoperative trials. They reported 79% doing well at 6 months and 73% doing well at 3 to 5 years. Similarly, Long suggested that prior to implantation, patients should have a defined organic basis for their pain and have had proper management of depression and other psychological problems.[24]

Of similar importance is pretesting for effectiveness. Methods vary from epidural steroid injections to TNS to temporary percutaneous monopolar stimulation;[18] in recent years, combinations of the three have frequently been used.[6,11,12,14,24]

The complications of spinal cord stimulation are largely mechanical. Subdural electrodes, CSF leaks, infection, and electrode migration have occasionally been seen.[32,54] Technical problems with percutaneous electrodes are more common. LeRoy reported a 14% rejection rate, an infection rate of 6%, and electrode migration in 6%.[19] Electrode migration of only 2 mm has been shown to cause complete loss of analgesia; attempts to prevent this problem have been unsuccessful. Lead breakage, presumably related to metal fatigue, is not uncommon, especially with the percutaneous technique. Indeed, it has been suggested that the higher complication rate seen with percutaneous techniques counterbalances the ease of insertion.[54]

The results of clinical investigations have been fairly uniform with approximately 50% of patients reporting good results, including increased activity and decreased narcotic use.[12-15,19,22,24,30,31,35,38] Young, on the other hand, reported poorer results, with 33% of patients discontinuing narcotics and 16% returning to work.[54] The lack of objective measures of improvement and the complexity of the pain experience make it difficult to precisely determine the effectiveness of any treatment for intractable pain. However, it is clear that spinal cord stimulation has become an important tool in the management of chronic pain.

ACKNOWLEDGMENT

This research was supported in part by the Veterans Administration Medical Center Research Grant #6000-02P.

REFERENCES

1. **Bantli, H., Bloedel, J. R., Long, D. M., et al.,** Distribution of activity in spinal pathways evoked by experimental dorsal column stimulation, *J. Neurosurg.*, 42, 290, 1975.
2. **Campbell, J.,** Examination of possible mechanisms by which stimulation of the spinal cord in man relieves pain, *Appl. Neurophysiol.*, 44, 181, 1981.
3. **Coburn, B.,** Electrical stimulation of the spinal cord: two-dimensional finite element analysis with particular reference to epidural electrodes, *Med. Biol. Eng. Comput.*, 18, 573, 1980.
4. **Cook, A. W. and Weinstein, P.,** Chronic dorsal column stimulation in multiple sclerosis: preliminary report, *N.Y. State J. Med.*, 73, 2868, 1973.
5. **Doerr, M., Krainick, J. U., and Thoden, U.,** Pain perception in man after long term spinal cord stimulation, *J. Neurol.*, 217(4), 261, 1978.
6. **Erickson, D. L.,** Percutaneous trial of stimulation for patient selection for implantable stimulating devices, *J. Neurosurg.*, 43, 440, 1975.
7. **Foreman, R. D., Beall, J. E., Applebaum, A. E., et al.,** Effects of dorsal column stimulation on primate spinothalamic tract neurons, *J. Neurophysiol.*, 39, 534, 1976.
8. **Gildenberg, P. L. and Murthy, K. S. K.,** Influence of dorsal column stimulation upon human thalamic somatosensory-evoked potentials, *Appl. Neurophysiol.*, 43, 8, 1980.
9. **Groth, K.,** personal communication.
10. **Hoppenstein, R.,** Electrical stimulation of the ventral and dorsal columns of the spinal cord for relief of chronic intractable pain: preliminary report, *Surg. Neurol.*, 4, 187, 1975.
11. **Hoppenstein, R.,** Percutaneous implantation of chronic spinal cord electrode for control of intractable pain, *Surg. Neurol.*, 4, 195, 1975.
12. **Hosobuchi, Y., Adams, J., and Weinstein, P. R.,** Preliminary percutaneous dorsal column stimulation prior to permanent implantation, *J. Neurosurg.*, 37, 242, 1972.
13. **Hosobuchi, Y., Ed.,** Indication for spinal cord stimulation, in *Proc. Symp. 7th Int. Cong. Neurosurgery*, Munich, July 12, 1981.
14. **Kasprak, M. E.,** Percutaneous stimulation in the spinal cord, *J. Neurosurg.*, 9(2), 67, 1977.
15. **Krainick, J., Thoden, U., and Riechert, T.,** Pain reduction in amputees by long-term spinal cord stimulation, *J. Neurosurg.*, 52, 346, 1980.
16. **Larson, S. J., Sances, A., Jr., Riegel, D. H., et al.,** Neurophysiological effects of dorsal column stimulation in man and monkey, *J. Neurosurg.*, 41(2), 217, 1974.
17. **Larson, S. J., Sances, A., Jr., Cusick, J. F., et al.,** A comparison between anterior and posterior spinal implant systems, *Surg. Neurol.*, 4, 180, 1975.
18. **LeRoy, P. L., Boulos, M. I., Monteleone, I. V.,** Percutaneous electrical stimulation (PES): a new modulation technique for the study of segmental afferent and efferent responses in humans, 1st World Cong. Pain, Florence, 1975.
19. **LeRoy, P. L.,** Stimulation of the spinal neuraxis by biocompatible electrical current in the human, *Appl. Neurophysiol.*, 44, 187, 1981.
20. **Long, D. M. and Hagfors, N.,** Electrical stimulation of the nervous system. The current status of electrical stimulation of the nervous system for relief of pain, *Pain*, 1, 109, 1975.
21. **Long, D. M.,** Electrical stimulation for the control of pain, *Arch. Surg.*, 112(7), 884, 1977.
22. **Long, D. M.,** Uses of percutaneous electrical stimulation of the nervous system, *Med. Prog. Technol.*, 5(1), 47, 1977.
23. **Long, D. M.,** Electrical stimulation of the nervous system for pain control, *Electroencephalogr. Clin. Neurophysiol. (Suppl.)*, 34, 343, 1978.
24. **Long, D. M., Erickson, D., Campbell, J., et al.,** Electrical stimulation of the spinal cord and peripheral nerves for pain control: a 10-year experience, *Appl. Neurophysiol.*, 44, 207, 1981.
25. **Melzack, R. and Wall, P. D.,** Pain mechanisms: a new theory, *Science*, 150, 972, 977, 1965.
26. **Myklebust, J., Sances, A., Jr., Swiontek, T., et al.,** Spinal cord stimulation studies, in *Proc. 6th Int. Symp. on Electrostimulation*, Albena, Bulgaria, September 24, 1981.
27. **Myklebust, J. B. and Cusick, J. F.,** Somatosensory evoked potentials in experimental neural stimulation, in *Neural Stimulation*, Vol. 1, Myklebust, J. B., Sances, A., Larson, S. J., and Cusick, J. F., Eds., CRC Press, Boca Raton, Fla., 1985, chap. 5.
28. **Nathan, P. W. and Rudge, P.,** Testing the gate control theory of pain in man, *J. Neurol. Neurosurg. Psychiatry*, 37, 1366, 1974.
29. **Nathan, P. W.,** The gate-control theory of pain: a critical review, *Brain*, 99, 123, 1976.
30. **North, R. B., Fischell, T. A., and Long, D. M.,** Chronic stimulation via percutaneously inserted epidural electrodes, *Neurosurgery*, 1(2), 215, 1977.
31. **North, R. B., Fischell, T. A., and Long, D. M.,** Chronic dorsal column stimulation via percutaneously inserted epidural electrodes. Preliminary results in 31 patients, *Appl. Neurophysiol.*, 40, 184, 1977-78.

32. **Pineda, A.,** Complications of dorsal column stimulation, *J. Neurosurg.*, 48, 64, 1978.
33. **Plotkin, R.,** Pisces stimulation for motor neurone disease, *Acta Neurochir. (Suppl.)*, 30, 429, 1980.
34. **Read, D. J., Matthews, W. B., and Higson, R. H.,** The effect of spinal cord stimulation on function in patients with multiple sclerosis, *Brain*, 104(4), 803, 1980.
35. **Richardson, R. R. and McLone, D. G.,** Percutaneous epidural neurostimulation for paraplegic spasticity, *Surg. Neurol.*, 9(3), 153, 1978.
36. **Richardson, R. R., Cerullo, L. J., and Meyer, P. R.,** Autonomic hyper-reflexia modulated by percutaneous epidural neurostimulation: a preliminary report, *Neurosurgery*, 4(6), 517, 1979.
37. **Richardson, R. R., Cerullo, L. J., McLone, D. G., Gutierrez, F. A., and Lewis, V.,** Percutaneous epidural neurostimulation in modulation of paraplegic spasticity. Six case reports, *Acta Neurochir.*, 49(3-4), 235, 1979.
38. **Richardson, R. R., Siquerira, E. B., and Cerullo, L. J.,** Spinal epidural neurostimulation for treatment of acute and chronic intractable pain: initial and long term results, *Neurosurgery*, 5(3), 344, 1979.
39. **Rusinko, J. B., Walker, C. F., and Sepulveda, N. G.,** Finite element modeling of potentials within the human thoracic spinal cord due to applied electrical stimulation, *IEEE 1981 Frontiers of Engineering in Health Care*, IEEE, Piscataway, N.J., 1981, 76.
40. **Sances, A., Jr. and Larson, S. J.,** *Electroanesthesia: Biomedical and Biophysical Studies*, Academic Press, New York, 1975.
41. **Sances, A., Jr., Swiontek, T. J., Larson, S. J., et al.,** Innovations in neurologic implant systems, *Med. Instrum.*, 9(5), 213, 1975.
42. **Sances, A., Jr., Weber, R., Myklebust, J., Cusick, J., Larson, S., and Saltzberg, B.,** The evoked potential: an experimental method for biomechanical analysis of brain and spinal injury, *SAE Trans.*, 89, 3815, 1980.
43. **Sedgwick, E. M., Thornton, A. R. D., El-Negamy, E., et al.,** Electrophysiological responses and stimulus parameters associated with spinal cord stimulation, *Int. Rehab. Med.*, 2(2), 71, 1980.
44. **Sepulveda, N. G. and Walker, C. F.,** Finite element modeling of potentials within the human brain due to applied electrical stimulation, *2nd IEEE-EMBS Conf.*, IEEE, Piscataway, N.J., 1980.
45. **Shealy, C. N., Mortimer, J. T., and Reswick, J. B.,** Electrical inhibition of pain by dorsal column stimulation. Preliminary clinical report, *Anesth. Analg. Curr. Res.*, 46, 489, 1967.
46. **Shimoji, K., Matsuki, M., Shimizu, H., et al.,** low frequency, weak extradural stimulation in the management of intractable pain, *Br. J. Anaesth.*, 49(11), 1081.
47. **Swiontek, T. J., Sances, A., Jr., Larson, S. J., et al.,** Spinal cord implant studies, *IEEE Trans. Biomed. Eng.*, 23(4), 307, 1976.
48. **Swiontek, T., Maiman, D., Sances, A., Jr., Myklebust, J., Larson, S. J., and Hemmy, D.,** Effect of electrical current on temperature and pH in cerebellum and spinal cord, *Surg. Neurol.*, 14(5), 365, 1980.
49. **Swiontek, T. and Sances, A., Jr.,** Experimental current density measurement, in *Neural Stimulation*, Vol. 1, Myklebust, J. B., Sances, A., Larson, S. J., and Cusick, J. F., Eds., CRC Press, Boca Raton, Fla., 1985, chap. 4.
50. **Torebjork, H. E. and Hallin, R. G.,** Responses in human A and C fibers to repeated electrical intradermal stimulation, *J. Neurol. Neurosurg. Psychiatry*, 37, 653, 1974.
51. **Urban, B. J. and Nashold, B. S., Jr.,** Percutaneous epidural stimulation of the spinal cord for relief of pain. Long-term results, *J. Neurosurg.*, 48(3), 323, 1978.
52. **Wall, P. D. and Sweet, W. H.,** Temporary abolition of pain in man on stimulation of large diameter cutaneous afferent fibers, *Science*, 155, 108, 1967.
53. **Wall, P. D.,** The gate control theory of pain mechanisms: a re-examination and re-statement, *Brain*, 101, 1, 1978.
54. **Young, R. F.,** Evaluation of dorsal column stimulation in the treatment of chronic pain, *Neurosurgery*, 3(3), 373, 1978.

INDEX

A

B

C

D

E

F

G

H

I

L

M

N

P

Q

R

S

T

U

V

W